ROUTLEDGE LIBRARY EDITIONS:
COMPARATIVE URBANIZATION

Volume 6

THE STRUCTURE OF NINETEENTH CENTURY CITIES

THE STRUCTURE OF NINETEENTH CENTURY CITIES

Edited by
JAMES H. JOHNSON AND COLIN G. POOLEY

Routledge
Taylor & Francis Group

LONDON AND NEW YORK

First published in 1982 by Croom Helm Ltd

This edition first published in 2021
by Routledge
2 Park Square, Milton Park, Abingdon, Oxon OX14 4RN

and by Routledge
52 Vanderbilt Avenue, New York, NY 10017

Routledge is an imprint of the Taylor & Francis Group, an informa business

British Library Cataloguing in Publication Data
A catalogue record for this book is available from the British Library

ISBN: 978-0-367-75717-5 (Set)
ISBN: 978-1-00-317423-3 (Set) (ebk)
ISBN: 978-0-367-77208-6 (Volume 6) (hbk)
ISBN: 978-0-367-77214-7 (Volume 6) (pbk)
ISBN: 978-1-00-317026-6 (Volume 6) (ebk)

Publisher's Note
The publisher has gone to great lengths to ensure the quality of this reprint but points out that some imperfections in the original copies may be apparent.

Disclaimer
The publisher has made every effort to trace copyright holders and would welcome correspondence from those they have been unable to trace.

THE STRUCTURE OF NINETEENTH CENTURY CITIES

Edited by
James H. Johnson
and Colin G. Pooley

CROOM HELM
London & Canberra

Croom Helm Ltd, 2-10 St John's Road, London SW11

British Library Cataloguing in Publication Data

The Structure of nineteenth-century cities.
  1. Cities and towns — Great Britain — History
  I. Johnson, J.H. II. Pooley, C.G.
  307.7'6'0941   HT133
  ISBN 0-7099-1412-1

First published in the United States of America in 1982
All rights reserved. For information write:
St. Martin's Press, Inc., 175 Fifth Avenue, New York, N.Y.10010

Library of Congress Cataloging in Publication Data

Main entry under title:
The Structure of nineteenth century cities.

  1. Cities and towns — Great Britain — History — 19th
century — Addresses, essays, lectures. I. Johnson,
James Henry, 1930-   . II. Pooley, C.G.
HT133.S85   1982   307.7'6'0941   81-21280
ISBN 0-312-76781-1   AACR2

Printed and bound in Great Britain by
Biddles Ltd, Guildford and King's Lynn

CONTENTS

TABLES AND FIGURES

Tables

Figures

PREFACE

These essays are the outcome of a series of seminars on the internal structure of the nineteenth-century British city, held in the Department of Geography at the University of Lancaster during 1978. Funding for the seminars was generously provided by the Social Science Research Council as part of its policy of encouraging interdisciplinary research through small-group meetings; and participants from a variety of disciplinary backgrounds were invited to prepare papers and contribute to discussions.

The idea for this seminar series grew from our concern that, although there were now large numbers of empirical studies of nineteenth-century urban history and historical geography, there had been little attempt to search for links between these studies and relate them to more general theories of nineteenth-century urban development. In order to further these aims the interdisciplinary seminar series was proposed. To ensure some overall coherence and to give the seminars a distinctive flavour, the topics chosen for discussion were all closely linked to the geographical study of the nineteenth-century city. Nonetheless they are also topics which have received considerable attention from non-geographical researchers on nineteenth-century towns, and hence they represent areas where cross-fertilization of ideas between disciplines is essential if significant advances are to be made.

Although the essays in this volume represent the majority of topics covered in the seminar series, there has been no attempt to cover comprehensively all aspects of the internal structure of the nineteenth-century British city. Indeed, a conscious decision was taken to study particular topics in some depth, rather than to include all possible aspects of nineteenth-century urban history.   The essays in this volume must thus be viewed only as one possible selection of important themes in British urban history.   The contributions have been grouped around three main themes: the physical structure of the city and the urban environment; occupations and the urban labour market; and the residential structure of nineteenth-century towns.   As far as is possible each section includes essays which reflect a variety of disciplinary approaches.

We thank our contributors for the prompt delivery of their carefully revised manuscripts and all participants in the seminar series who took part in the discussions with vigour and good humour.   The support of the Social Science Research Council converted our original idea into reality, and we are grateful for its backing.   The secretarial and technical staffs in the Department of Geography at Lancaster University have helped in many ways throughout the project with unfailing cheerfulness and we are particularly grateful to Mrs. Christine Skinner for typing the final version for publication, to Mrs. Anne Jackson for her help with illustrations and to Mrs. Marian Jackson for her assistance with the index.   Our greatest debt of all is to the large number of researchers who, over the past three decades, have provided the diverse body of work which made the seminar series both desirable and possible.   If the essays in this volume make some small contribution towards stimulating further good research into aspects of nineteenth-century urban history and geography, then this book will have fulfilled its primary aim.

University of Lancaster   James H. Johnson
                                           Colin G. Pooley

PART ONE

INTRODUCTION

Chapter 1

THE INTERNAL STRUCTURE OF THE NINETEENTH-CENTURY BRITISH CITY - AN OVERVIEW

James H.Johnson and Colin G.Pooley

The development of nineteenth-century urban research

The study of nineteenth-century cities has been one of the most rapidly expanding areas of British historical research in recent years. Review articles and research registers clearly show that during the last two decades nineteenth-century urban studies have moved from a position of comparative neglect to one almost of dominance (Dyos, 1968; Wild, 1980; Urban History Year-Book). Victorian commentators were well aware of the rapid and extensive urban growth which was taking place around them, and they produced many descriptive and statistical accounts of nineteenth-century urbanization (Vaughan, 1843; Engels, 1845; Booth, 1889; Weber, 1899). Modern interest in nineteenth-century towns, however, stems from the 1950s and 60s, with the beginnings of the present surge of activity being based on seminal studies by scholars such as Chaloner (1950), Barker and Harris (1954), Lawton (1955), Dyos (1961) and Briggs (1963). Dyos (1966) has reviewed a number of important early analyses of nineteenth century towns, but since he wrote the development of urban historical research has been even more rapid with the publication of several major books (Anderson, 1971; Stedman-Jones, 1971; Armstrong, 1974; Foster, 1974; Fraser, 1976; Daunton, 1977; Crossick, 1978) together with many important articles and numerous theses (Dyos and Wolff, 1973; Whitehand and Patten, 1977; Dennis, 1979).

Probably the most significant recent development for British urban history was the founding of the Urban History Group in 1963, under the energetic leadership of the late Professor H.J. Dyos, leading to the publication of the proceedings of the 1966 Urban History Group meeting (Dyos, 1968). In 1966 Dyos was able to write 'by contrast with the torpid interest in the subject before the war the volume of writing on the history

3

of British towns now looks like a cataract'.
Successive meetings of the Group have ensured that
this research interest in nineteenth-century towns
has continued unabated.

The upsurge of interest in the nineteenth-
century city has not been confined to the United
Kingdom, and one of the most productive strands of
recent urban historical research has been the
cross-fertilization of ideas between Britain and
Europe, Australasia and, in particular, North
America.   The North American interest in nineteenth-century urban development emerged rather
earlier and grew even more quickly than in Britain.
By the early 1960s several major works on urban
history had appeared (Handlin and Burchard, 1962;
Glaab, 1963; Thernstrom, 1964; Hauser and Schnore,
1965).   Each of these drew on earlier experiences
and, most importantly, pointed the way forward to
new and substantial studies including those by
Goheen (1970), Knights (1971), Thernstrom (1973),
Katz (1976) and Conzen (1976).   Similarity of data
sources and methods of analysis has meant that
there is considerable scope for comparisons between
British and American towns in the nineteenth
century, although relatively few scholars have
specifically tackled studies which involve trans-
atlantic comparisons (Ward, 1964, 1974, 1975;
Sharpless, 1976).

Elsewhere, developments in nineteenth-century
urban history have been less rapid, but a substantial body of literature on the development of
European and Antipodean towns is now emerging
(Sutcliffe, 1970; Chevalier, 1973; Bater, 1976;
Ogden, 1977; Lawson, 1973; Williams, 1974).   Comparative studies would certainly aid our understanding of the processes of nineteenth-century
urbanization in spite of detailed contrasts in the
time and context of urban development, but such
studies are few (Lees, 1973).

As well as the wide geographical spread of
recent studies, urban historical research has also
developed a strong interdisciplinary flavour, since
most topics in urban history can be approached from
a variety of viewpoints.   As a result there are
almost as many studies by geographers, sociologists
and political scientists as have come from the pens
of historians.   Sometimes the disciplinary backgrounds of individual authors are almost indis-
tinguishable in their writings, but there are a
number of areas where specialists from different
fields have devoted the major part of their
research efforts, whilst differences in method and

4

technique are also sometimes apparent.

Social and economic historians have concentrated most fully on the influence of class and ideology on the structure of nineteenth-century towns (Stedman-Jones, 1971; Neale, 1972; Gray, 1976; Crossick, 1976), on the effects of municipal policy and reform on subsequent urban structure (Sutcliffe,1970; Hennock, 1973; Kellett, 1978),on the related problems of health and housing in nineteenth-century towns (Chapman, 1971; Gauldie, 1974; Wohl, 1977), and on problems of the economic development of cities and the histories of individual trades (Hyde, 1971; Anderson, 1976; Floud, 1976). Although not their exclusive preserve, such topics have become dominated by the work of historians and are characterized by the careful evaluation of a wide range of disparate sources concerning the political, social and economic development of the city.

In contrast the work of geographers has tended to concentrate on the spatial aspects of nineteenth-century urban development, often focused on a smaller range of sources such as the census, cartographic evidence and other data which facilitate spatial referencing. A number of topics particularly lend themselves to this approach and thus have attracted geographical research. Studies of urban physical morphology perhaps best exemplify the spatial approach (Conzen, 1960; Carter, 1965; Whitehand, 1978; Fox, 1979), but in recent years this has been more generally extended to the identification of social areas in cities. In most cases these studies have involved the extension of current urban geographical theories and techniques back to the nineteenth-century town, leading to analyses of socio-spatial variation, based mainly on the census enumerators' books (Warnes, 1972; Lawton and Pooley, 1976; Shaw, 1977; Cowlard, 1979), but also using other sources such as urban rate books (Gordon, 1979). As well as the identification of broad social areas, geographers have also concentrated on particular aspects of urban demography and social structure, including rural-urban migration, the formation of immigrant areas in cities (Lawton, 1958; Ward, 1971; Pooley, 1977), and the processes of residential mobility and spatial change in urban areas (Pritchard, 1976; Dennis, 1977b; Pooley, 1979a). Again, although not the sole preserve of geographers, spatial studies of nineteenth-century towns by workers from other disciplines are relatively uncommon (Cannadine, 1977; Daunton, 1977).

Work by sociologists on the nineteenth century city has expanded less rapidly than studies by historians and geographers, but there are a number of areas where a distinctive sociological contribution can be discerned. Family structure and the role of the family in the process of urbanization is one important area of sociological interest (Smelser, 1959; Anderson, 1971, 1976), whilst the study of kinship patterns and neighbourhood communities in nineteenth-century towns involves a combination of geographical and sociological skills (Pahl, 1970; Dennis, 1977a; Macfarlane, 1977a). Much of this work has been carried out at a micro-scale, involving the reconstruction of household and family characteristics, although the detailed historical-anthropological work which has been attempted for earlier periods has yet to be properly extended to the nineteenth century (Macfarlane, 1977b). One of the aims of the seminar series from which the essays in this volume are derived was to explore the distinctive approaches of different researchers to the study of the internal structure of the nineteenth-century city. Undoubtedly most good research draws inspiration from a range of disciplinary backgrounds, but as contributions to this volume demonstrate, there remain important differences in approaches and methods of analysis. It would seem that there is still much to be gained from the cross-fertilization of ideas between scholars from different disciplines.

Why study the nineteenth-century city?

The sudden upsurge of interest in the study of nineteenth-century towns is difficult to explain. Some of this activity may be attributed to a 'snow-ball' effect following the first few major studies and the formation of the Urban History Group, but there are three substantive reasons why nineteenth-century urban research is important today and should remain so in the future.

First, the process of nineteenth-century urban development has much wider implications. Industrialization and urbanization in the nineteenth century brought with them fundamental changes in the spatial organisation of towns and in the relationship between classes and groups within urban society. Such changes have occurred in all developed nations at some time in the recent past, while related processes are still operating in developing nations today. The study of nineteenth-century cities can thus lead into attempts to generate more generally-applicable theories about the processes of modernization, urbanization and associated social changes (Reissman, 1964; Germani, 1969).

Second, not only may nineteenth-century urban experiences be usefully compared with the process of urbanization in the developing world today (McGee, 1971; Friedman and Wulff, 1976), but the effects of nineteenth-century growth are still being felt in urban areas in Britain and Europe. Recent inner-area studies of British cities (Department of Environment, 1977a, 1977b) highlight the constraints which nineteenth-century housing stocks, road systems and public utilities place on modern urban regeneration, whilst the processes of suburbanization and inner-city decay have their roots firmly in the nineteenth-century industrial city. A greater understanding of the nineteenth-century city helps us to appreciate more fully the complex mechanisms operating within the modern metropolis.

Finally, scholars have also been attracted to the nineteenth-century city because of the fairly-ready availability of comprehensive data sources. In comparison with earlier periods, the mass of social, economic, demographic and political information which is available for nineteenth-century towns is almost overwhelming. This information is available at a variety of scales, ranging from the individual to large groups and whole urban populations. It is thus possible to examine the most complex processes operating within urban society and to apply theories culled from modern urban research. Indeed, because it is no longer necessary to preserve confidentiality, the volume of individually-based information is often better for the mid-nineteenth century than it is for the more recent past. This fact is undoubtedly significant in explaining the upsurge of interest in nineteenth-century towns, particularly focusing on the period 1841-1881 for which detailed census enumerators' returns are available. This mass of factual data also allows the reinterpretation of contemporary accounts of the process of urbanization, since their accuracy can be checked against precise statistical information which can often be analysed in spatial terms.

Data and method

Unusually for historical researchers, the student of the nineteenth-century town is often faced with a surfeit of data, although this information may not necessarily be in the form most useful for analysis or may not refer to the most relevant dates. Details of the diverse range of sources used in the studies reported here are given in individual essays, but some more general comments on source availability can be made here.

Sources for the study of nineteenth-century urban history can be broadly divided into two categories. Some sources provide a more-or-less systematic and comprehensive coverage of a topic in a form which permits quantification and statistical analysis on a fairly objective and rigorous basis. Other sources are much more sporadic in their spatial and temporal occurrence, and contain more subjective, but important, descriptive material. Inevitably a few sources will fall some way between these two ideal types. Examples of each are given in Table 1.1.

An upsurge of interest in nineteenth-century towns coincided with the so-called 'quantitative revolution' in the social sciences. Historians were slowest to move from their traditionally idiographic approach and embrace the new methodologies, but sociologists, geographers and economists rapidly appreciated that many of the techniques and methods of analysis, often borrowed from the physical and natural sciences, could be applied to some of the more objective, systematic and quantifiable data sources from the nineteenth century. The implications of this movement towards a more positivistic approach to social science have been the subject of considerable discussion (Keat and Urry, 1975), but perhaps three assumptions can be seen as central to these changes (King, 1976). First, the positivistic approach assumes that the methodological procedures of natural science may be directly adopted by the social sciences, in particular, that regardless of the influence of the free-will of individual actors, the behaviour of objects under study may be observed, measured and tested against a prior hypothesis. Secondly, it is assumed that the outcomes of social investigations can be formulated in the same manner as in the natural sciences, that is that the aim of analysis is to derive theories and laws which have some general applicability.

8

Table 1.1: Major sources for the study of nineteenth-century urban society

	Objective	$\longrightarrow$	Descriptive
Comprehensive	Published census volumes from 1801	Medical Officer of Health Reports	Council minutes
	Census enumerators' books 1841–1881*		
	Vital Registration data from 1837	Parliamentary reports and enquiries	Newspapers
	Parish registers+		
	City directories°	Company papers and accounts	Contemporary comments
	Rate books+		
	Electoral rolls/poll books+		
	Maps	Poor Law, Police, Education reports	Miscellaneous archival material, personal diaries
Sporadic	Building registers and plans+		

* Subject to 100-year confidentiality constraint

o Variable quality and coverage

+ May not survive for some towns

Thirdly, it is assumed that investigations are objective and will thus generate information which is 'neutral' or 'value-free' in character.

Each of these tenets has been the subject of much criticism (Guelke, 1971; Harvey, 1973; Gregory, 1978), but it was inevitable that social scientists who accepted even part of the positivistic philosophy of investigation would be drawn towards those sources which provided large-scale sets of apparently systematic and objective data which could be used to test hypotheses and generate theories about nineteenth-century urban society. This philosophical revolution coincided with the technical changes that made computer facilities readily available to social-science researchers and as a result for the first time large data sets of census and other material could be handled with ease. Although urban historical researchers continued to use descriptive sources, in many cases they were regarded as secondary - and perhaps inferior - to the more objective data sets, particularly the census enumerators' books. As a result a very large number of local studies were undertaken, based on the manuscript census returns. Hence, too, a number of textbooks were published describing these sources and discussing methods of classification, coding and computerization (Wrigley, 1966, 1972; Lawton, 1978).

Pioneering work on the census enumerators' books introduced more rigorous modes of analysis (Anderson, 1971; Armstrong, 1974) and provided an invaluable jolt to urban historical research, which set in motion a number of increasingly ambitious projects based primarily on censuses, rate books and other quantifiable data. The resultant studies have produced the most comprehensive and detailed information yet available about nineteenth-century towns, but there are also dangers in this approach. First, it can be argued that the philosophical assumptions of positivism are often inappropriate to the study of any social science and the associated methodological procedures should thus be used with caution. Second, it may be suggested that even if a positivistic approach is accepted, concentration on the quantitative analysis of large data sets has led to a lack of careful selection of theories to be tested or problems to be investigated and, in particular, to the relative neglect of more traditional descriptive historical sources. Each of these criticisms may be briefly assessed.

Most basic is the argument that the methods of the natural sciences are inapplicable, because social scientists are in the last resort dealing with the behaviour of actors with individual free-will. As a result the social and economic processes operating within urban areas are not analogous to a series of objects behaving in a predictable way which can be objectively observed, measured, tested and from which generally-applicable laws can be generated. Definite laws are thus difficult to formulate and the strictly-defined scientific method is inappropriate. Furthermore, the social scientist can never be totally objective in his analysis of society, as a 'value-free' judgement implies complete detachment from the objects with which he is dealing. Any researcher must be conditioned by his experience of the society in which he lives and thus the interpretation of processes operating in urban society, past or present, must contain some element of the researcher's own subjective judgement.

Such criticisms of a positivistic approach to the social sciences are elaborated in many texts (Keat and Urry, 1975; Gregory, 1978). From the point of view of practical research into nineteenth-century urbanism, however, something may be salvaged from the debate. Whilst total objectivity of interpretation is clearly impossible, it is feasible to adopt a somewhat more detached view of a past society, even though final judgements and interpretations may be value-laden. Furthermore, analysis of the large data sets which are available for the mid-Victorian period demands a certain degree of rigour. Although strictly scientific laws cannot be derived, hypotheses can be tested and the general tendencies which operate at the aggregate scale within society can be identified. Such an approach may only go some way towards total explanation and a truly positivistic explanation will never be attainable, but there seems no reason why a 'scientific method' should not be applied to the study of past society to provide one important strand among others in the cluster of approaches that are inevitably needed for total explanation.

The second criticism can be examined more conclusively, as this involves a more straight-forward assessment of the practical approaches most appropriate to the study of urban history. Although a positivistic approach to urban historical research does not necessarily mean that the study of individuals and small groups will be neglected, in practice it has tended to lead to the use of aggregate statistics which blur the infinite variety of individual behaviour, in an attempt to generate more widely-applicable theories. It can be suggested that too many studies have adopted an all-encompassing approach, attempting to describe, at an aggregate level, all aspects of mid-Victorian society, but paying little attention to the selection of significant historical problems, to the role of the individual in fashioning Victorian society or to the development of theory directly relevant to nineteenth-century urban society.

These criticisms are largely valid, particularly when applied to some of the large-scale census studies carried out by geographers and some social historians. Such studies were probably necessary in order to lay the foundations for future research, but it is now time for nineteenth-century urban historical research to become more selective in the themes which it tackles, attempting to answer important historical questions through in-depth analyses, rather than providing blanket descriptions of all aspects of Victorian society. The essays and discussion reported in this volume are an attempt to define what these important historical questions may be. More detailed and selective studies will also inevitably involve movement away from the generation of theory from aggregate data, towards a much fuller consideration of individual actions and behaviour. This does not mean that a scientific methodology is necessarily inappropriate, since it can be applied equally effectively at the individual as well as at the aggregate level. What it does imply is that greater account must be taken of the variety of individual responses that make up aggregate patterns, and that the underlying actions which constitute aggregate processes need to be more fully investigated. Geographers have been aware for some time that ecological analysis tells us little about individual behaviour and that aggregate spatial analysis must be followed up at the individual and small-group level (Robinson, 1950; Duncan, Cuzzort and Duncan, 1961; Dogan and Rokkan, 1969). This knowledge has not always been carried through effectively to urban historical research. Historians, on the other hand, have traditionally been more concerned with the small-scale study and have only reluctantly adopted an aggregate approach. Each approach is, in reality, complimentary to the other, but the recent over-emphasis on large-scale aggregate studies must be rectified if underlying processes and individual reactions are to be fully understood.

Although there is an abundance of descriptive and statistical data on the nineteenth-century city,

12

reliance on census enumerators' books has also constrained the way in which temporal change has been studied. Most frequently change over time has been inferred from a series of static cross-sectional descriptions of the city as it was on census night, with little attempt to study directly the processes of change which operated between census dates. Although complete time-series of data are rare, it is possible to link cross-sectional census analysis with the examination of sources such as directories and rate books which occur more often, and to fill in other gaps with descriptive material in order to learn more about the processes of change over time. Inevitably, too, source availability has encouraged researchers to concentrate on the middle years of the nineteenth century to the relative neglect of earlier and later periods. Although sources are less abundant and comprehensive for the periods before 1841 and after 1881, there is still much that can be discovered about these crucial years of change in urban society - a theme which is explored in several essays in this volume.

Themes in the study of the nineteenth-century city

In attempting to classify recent studies, division can be made along two main axes. First, there are those studies which are concerned with inter-urban comparisons, focusing on the growth of the urban hierarchy in the nineteenth century (Robson, 1973; Carter, 1978). Second, there are those which study variations within towns. Essays in this volume are concerned only with the second approach, the internal structure of nineteenth-century towns, both because of limitations of space and because the main thrust of research on nineteenth-century urbanism has recently been in this direction.

Research can also be divided between those studies which seek to relate a comprehensive history of particular towns and those which have a more thematic focus, studying particular topics with respect to one or more nineteenth-century cities. Although there are many invaluable histories of individual cities in the catalogue of urban research (Briggs and Gill, 1952; Barker and Harris, 1954; Hill, 1975), most recent studies have more often focused on one or more systematic themes in the history of urbanization. Thus, there have recently been studies of topics like immigrant groups, urban social differentiation, estate development and retail provision (for instance,

Lees, 1969; Pooley, 1977; Cannadine, 1977; Whitehand, 1975; Wild and Shaw, 1974). This is the approach adopted by essays in this volume, which collectively seek to advance our understanding of selected themes in the study of nineteenth-century urban structure.

The physical structure of the city

Studies of the physical structure of nineteenth-century cities can be broadly divided between three approaches. Geographers have often adopted a morphological approach in which the study of the physical form of the city assumes importance in its own right. The analysis of town plans was pioneered by Conzen (1960, 1966) and has since been attempted for several other towns (Carter, 1965; Straw, 1967). Though instructive, such approaches tend to be descriptive rather than analytical and a number of attempts have been made to extend the study of urban morphology into a fuller appreciation of the physical aspects of nineteenth-century urban development. The influence of previous land-ownership and boundary patterns on the developing urban form has been one line of approach (Ward, 1962; Mortimore, 1969), while this interest has been extended through concentration on the form of the urban fringe belt and nineteenth-century suburban growth (Whitehand, 1967, 1974, 1975). Perhaps most promising are recent attempts to link physical and social space in nineteenth-century towns by studying the relationships among urban morphology, land-use and emerging social areas (Vance, 1967; Fox, 1979; Carter and Wheatley, 1979; Shaw, 1979). These studies show that it is much more fruitful to view urban morphology as one of a series of interlinked aspects of the evolving urban structure rather than to study the physical development of urban areas through town-plan analysis alone. Just as the morphological character influenced land-use and social development, so the social and economic character of the city influenced subsequent physical growth. The further study of the interaction among these factors is one research frontier towards which studies of urban morphology should be advancing.

There are also a large number of studies of other aspects of the physical structure of the city, not necessarily directly related to morphological analysis. Recent research has focused on both the slums and the suburbs (Dyos, 1961, 1967; Tarn, 1971; Chapman, 1971; Dyos and Reeder, 1973; Chalklin, 1974;

14

Gauldie, 1974; Simpson and Lloyd, 1977; Burnett, 1978). Each in its different way typifies urban development in Victorian England, although most contemporary comment focused on the problems of the slums, while the bulk of new building was inevitably suburban. A number of carefully-researched studies deal with the development of particular estates (Prince, 1964; Thompson, 1974; Olsen, 1976; Daunton, 1977; Cannadine, 1977) and although the contribution of local authority and charitable trust housing to the total housing problem was small, this sector has also attracted a considerable amount of attention (Tarn, 1973; Sutcliffe, 1974).

Whilst these studies have concentrated mainly on the social history of housing, the development of the building industry and the mechanics of estate development, a number of more specifically economic studies have attempted to link the nature and timing of building to fluctuations in national and local economic fortunes. The work of Alonso (1964) and Parry-Lewis (1965) provided the basis for many current studies of urban land-values and building development, while these themes have been taken further in a series of studies by Whitehand (1972, 1975) and through interesting local case-studies, such as that by Forster (1972). Recently, Whitehand (1977) has attempted to place the study of nineteenth-century urban development within a sounder theoretical framework. Two essays in this book also deal with these themes: Aspinall concentrates on the structure of the building industry and its effects on urban growth, and Rodger focuses on the economics of house construction and the relationship of building activity to other economic fluctuations.

Although most new building took place on the periphery of Victorian cities, the housing that caused contemporary concern was the older central housing-stock which had filtered down to the urban poor and had become multi-occupied. As a result a further approach to the study of the physical fabric of the city has focused particularly on the slums, and is concerned with the quality of housing, the availability of sanitation and the impact of disease on urban areas. There are a number of general reviews of the impact of disease and the development of the sanitary movement in Victorian England (Rosen, 1971, 1973; Woodward and Richards, 1977; Smith, 1979), but relatively few studies consider the interaction of housing, disease and the urban environment in the context of specific towns. Wohl (1977) traces the development of housing legislation in London during the nineteenth century, Taylor (1976) looks specifically at the interaction of these factors in Liverpool, while a number of other studies review different aspects of the health and housing problem (Chapman, 1971; Gauldie, 1974; Woods, 1978). Arising out of contemporary nineteenth-century interest in the health of towns (for instance, Kay-Shuttleworth, 1832; Chadwick, 1842; Creighton, 1894), the later nineteenth century and early-twentieth century saw the development of local and national intervention in housing policy and the origins of modern town planning (Howard, 1898; Dewsnup, 1907; Geddes, 1915). Reasons for the growth of the town planning movement and its effectiveness in combating the worst evils of Victorian urban structure have themselves been the focus of several modern studies (Ashworth, 1954; Cherry, 1972, 1979; Sutcliffe, 1981; Yelling, 1981), while the administrative and political framework within which housing, health and sanitary legislation was formulated has also been extensively studied (Lewis, 1952; Brockington, 1965; Flinn, 1965; Hennock, 1973; Fraser, 1979). Sutcliffe's essay in this volume draws together a number of these strands, highlighting the physical and social problems attendant on Victorian urban growth and the ways in which faltering steps were made towards local and national intervention in housing and in the provision of basic public utilities.

Perhaps future research should move towards a greater integration of these various approaches to the study of the physical structure of the city and, in particular, should show greater interest in the operation of the housing market in the nineteenth century. A considerable amount is now known about the way in which cities grew, the factors affecting different types of development and, indeed, what it was like to live in different sorts of housing; but we still know relatively little of the way in which individuals gained access to different sectors of the housing market, the efficiency with which housing filtered from one sector of the population to another or the extent to which supply responded to demand in the different sub-markets of the housing system. Such research requires knowledge of the construction industry, economic and other factors affecting estate development, and the social and financial constraints on house purchase or rent. It can probably only be effectively undertaken for a few towns with a good range of documentary sources, but the returns on effort invested in such a project would be considerable.

The urban economy

Studies of the nineteenth-century urban economy have adopted a number of different approaches to the subject, although there is a noticeable neglect of the urban economy in geographical studies, which have focused instead on physical and social variations in nineteenth-century towns. Although not within the scope of this volume, the first, and most wide-ranging approach, is the macro-economic study of developing urban economies. Such studies range from attempts to chart economic and urban development over time (Rowstow, 1960; Dean and Cole, 1962; Thomas, 1972) to studies of the economic development of individual towns (Chaloner, 1950; Pollard, 1959; Church, 1966; Hyde, 1971) which often also include detailed studies of particular occupational sectors, such as Barker and Harris (1954) on glass-making in St. Helens.

One of the most significant results of urbanization is the associated change in occupational structure and the growth of distinctively urban employment. Many studies have thus concentrated on particular sectors of urban employment, which usually provide more manageable foci for study than entire towns. Transport development, for instance, was closely related to the history of urban growth and also provided a significant source of employment in nineteenth-century towns. The impact of the railway on employment opportunities (Patmore, 1962; Perkin, 1970; Kingsford, 1970) was both primary and secondary as not only were there jobs for railwaymen, but the coming of rail communications also frequently stimulated other areas of employment, as in the growth of resort towns (Gilbert, 1954; Walton, 1974). Elsewhere, in ports such as London and Liverpool, one of the most rapidly expanding employment sectors was dock and dock-related occupations (Lovell, 1969; Cunningham, 1970; Stedman-Jones, 1971; Taplin, 1974) while this sector was also well-documented in the late-nineteenth and early-twentieth century (Rathbone, 1904).

Urban-based commercial and service employment also became gradually more important during the nineteenth century, although its precise impact and the role which it played in the urban labour market has been only partly researched (Anderson, 1976; Crossick, 1977), while the development of urban retailing - a long neglected field - has recently

17

attracted rather more attention (Jeffreys, 1954; Alexander, 1970; Scola, 1975; Shaw and Wild, 1979). Manufacturing industry is the best documented of the distinctively urban occupations, with studies of the development of particular industries as well as histories of individual businesses and trade unions providing a more comprehensive coverage (Supple, 1977). Not surprisingly, the leading industrial sectors involved in the industrial revolution have received the fullest treatment, with the textile industry taking pride of place (Chapman, 1972; Gattrell, 1977; Farnie, 1979), closely followed by iron and steel (Ashworth; 1954; Daff, 1973), coal (Benson and Neville, 1976) and various branches of the rapidly-expanding engineering industries (Musson and Robinson, 1959; Smith, 1975; Floud, 1976). Although such trades and individual firms have been extensively studied (and most were predominantly urban-based in the nineteenth century), relatively few researchers set out to assess specifically the impact of these industries on urban development. While sociological studies of working conditions and the influence of factory production on home-life and traditional communities do exist (Hewitt, 1958; Smelser, 1959; Bythell, 1969), the influence of industrial development on the physical form of the city and on the nature of subsequent urban development seems to be a relatively neglected topic.

An alternative approach to the study of the urban economy shifts attention from individual industries and firms to the operation of the labour market, the way in which different groups in society gained access to work and the conditions of employment in different trades. This area of study is also relatively neglected, with little information available at present on the ways in which firms recruited their labour forces and on the career prospects in different occupations. Recent studies which have gone some way towards investigating these areas include Smelser (1959), Hobsbawm (1964), Stedman-Jones (1971) and Anderson (1976), while in this volume Green reviews the casual labour market in Victorian London with particular emphasis on the impact of street trading. It is clearly not possible for essays in a volume such as this to review the developing and highly-complex urban economy in its entirety, so the other two essays in this section are also highly focused, concentrating on the formal sector of the urban retail trade. From the viewpoints of historian and geographer, Scola and Shaw review the

18

relatively small number of studies that have been
completed so far and draw attention to the most
important changes which occurred in urban retailing
during the nineteenth century.

A final way in which the urban economy may be
studied is through the investigation of the inter-
relationship between urban form and economic and
technical changes during the nineteenth century.
The internal structure of the city was clearly
fashioned by a number of important influences, but
the direct effects of industry and technology are a
neglected field in historical research. As Shaw
points out, the development of multiple stores had
a considerable effect not just on shopping patterns
and trade areas, but also on the form of the
central business district. There have, however,
been few studies of the factors affecting the
development of British city centres to match those
for American towns (Ward, 1966; Bowden, 1977).
Industrial location decisions, together with other
types of urban redevelopment, also severely
affected subsequent urban growth, but such themes
are only rarely mentioned in studies of British
cities (Whitehand, 1978; Carter and Wheatley, 1979)
and even the effects of transport developments on
urban form and subsequent growth have been
curiously neglected since Kellett's (1969) pioneer
work.

Thus although there are large numbers of
business histories, economic histories of individual
trades and macro-level studies of the nineteenth-
century economy, the relationship between economic
development, the operation of the labour market and
the internal structure of the city is much more
rarely assessed. It must not be forgotten that
factory employment and the development of urban
occupations had not only social and economic but
also spatial implications for urban life. In this
vast area of study there is considerable scope for
further co-operation between geographers and
economic historians.

Urban social structure

Differences of approach to the study of urban
social structure can be typified by the various
scales at which analysis is attempted. Geographers
have traditionally placed most emphasis on an
'ecological' approach, in which the researcher
seeks to identify reasonable large and socially-
homogeneous areas through the amalgamation of small
spatial units, such as census enumeration districts,

19

according to the social characteristics which they
possess.   Although early work used subjective
criteria to define social areas (Lawton, 1955), most
recent studies have adopted multivariate data
analysis from modern urban geography in attempts to
define the most significant dimensions in Victorian
urban structure, grouping them to create object-
ively-defined social areas (Warnes, 1972; Tansey,
1973; Lawton and Pooley, 1975b; Shaw, 1977; Carter
and Wheatley, 1978a, 1978b).   Other studies have
concentrated on single variables such as rateable
value (Gordon, 1979) or a composite measure of
social status (Cowlard, 1979), but in each case the
initial (though not necessarily the ultimate) aim
has been to describe the social ecology of the city
in aggregate terms.   Relatively few studies by
historians have adopted an explicitly spatial
approach and this remains one important area where
there is much scope for cross-fertilization of
ideas and methodologies between geographers and
historians.   One major problem with the
'ecological' approach, however, is that as its
data are largely census-based it is difficult to
study processes of change in social areas over
time.   Some attempt to overcome this problem has
been made in recent studies of the rate and pattern
of residential mobility in Victorian towns and,
where knowledge of residential mobility can be
related to a socially-defined spatial base, the
manner in which social areas changed their
character can be interpreted (Pritchard, 1976;
Dennis, 1977b; Daunton, 1977; Pooley, 1979a).

Developments from the traditional 'social-area'
approach to the study of other aspects of urban
social structure have been in two directions,
either towards the aggregate study of specific
social groups in society, or towards the analysis
of small-scale neighbourhood communities, families
and even individuals.   Specific groups which
deserve more detailed study have been selected both
on subjective grounds and following multivariate
analysis of large data sets, but in both cases the
group most often singled out for attention is the
Irish.   There are now case-studies of the location,
community organization and familial and social
characteristics of the Irish in a number of towns
(Richardson, 1968; Lees, 1969, 1979; Lobban, 1971;
Werly, 1973; Dillon, 1974), but there are as yet
relatively few studies of other migrant groups with
which the Irish may be compared (Lipman, 1954;
Williams, 1976; Pooley, 1977).   Equally, the
residential location and social organization of

specific occupational groups has received only passing mention in most studies (Lawton and Pooley, 1976; Armstrong, 1974; Carter and Wheatley, 1978c; Ward, 1980). Although such studies are beginning to delve beneath the surface of aggregate social description, there is still much to be discovered about the ways in which specific groups were socially and spatially organized in nineteenth-century towns.

Dissatisfaction with the aggregate approach has led several researchers into studies of very small neighbourhoods, of the family and of individual behaviour, both as research topics in their own right and as illustrations of the micro-scale processes which collectively form the more easily-described aggregate patterns. A major problem of research into urban social structure at this historical-anthropological level is the lack of direct information on the behaviour of individuals in the past. Anderson (1971) and Dennis (1977a) have attempted studies of the family and of social networks using a combination of sources, including the census, vital registration certificates and a host of archival material. Such detailed research inevitably involves a very large investment of effort in record linkage, but this is essential if large numbers of family histories and case studies are to be reconstructed. Other sources such as diaries, personal reminiscences and, for the late-nineteenth and early-twentieth century, oral history all provide invaluable additional information; but the major problem of interpreting such reports lies in the assessment of their typicality. Oral history projects where large numbers of residents are interviewed provide the best chance of gaining a cross-section of society (Thompson, 1973; Roberts, 1976), but personal accounts and diaries are not only highly subjective but probably only survive for the least typical members of society (Roberts, 1971; Burnett, 1974; Lawton and Pooley, 1975a). Nevertheless, despite these difficulties, the probing of small-scale communities, individual reactions and personal behaviour is an essential complimentary approach to macro-scale ecological studies.

Whilst geographers inevitably emphasize the spatial aspects of urban social structure, studies by social historians have tended to focus instead on the network of social relations and the class structure which developed within the nineteenth-century town. Wide-ranging reviews of the emerging class structure in English society (Thompson, 1963;

Perkin, 1969) have been followed up by a number of detailed case studies of particular towns and groups within society (Stedman-Jones, 1971; Foster, 1974; Gray, 1976; Crossick, 1978; Morris, 1979). Although each of these studies takes a slightly different approach to the problem of measuring 'class' and 'status', and interpretations are inevitably affected by the political and ideological framework within which analysis is couched, each contributes significantly to our knowledge about the ways in which social classes were formed and perceived in the nineteenth century.

Receiving rather less attention in most of these studies, however, is the extent to which spatial segregation and social segregation were linked, a theme specifically developed by Cannadine in this volume. Other essays in this section show a progression through the scales of analysis mentioned above: Pooley is mainly concerned with the process of residential differentiation at the aggregate level while Dennis reviews ways in which smaller neighbourhood communities may be investigated. Anderson, focusing particularly on change in urban communities, sounds a cautionary note about studies of individual residential mobility as indicators of change in urban social areas. Clearly any of the above approaches requires a large investment of time and effort in data collection. Before embarking on such studies it is important to establish the precise questions to be asked. It is also important that the approach adopted is one which is likely to provide an answer to these questions. Whilst aggregate and individual approaches are both valid, some questions can only be answered at an individual level, whilst others respond best to an aggregate approach. A greater flexibility of approach to the study of urban social structure may be appropriate for future research in this field.

Conclusion

It is clear that research into urban history during the last two decades has provided us with a wide range of descriptive and analytical material on most aspects of urban life. We now know what the internal structure of the city was like in the mid-nineteenth century, in its physical, economic and social manifestations. The desire to collect factual information about Victorian towns has led to a large number of detailed case studies, but a relative neglect of the theories of urban development which are needed to draw this research together and to place the different case studies in their proper context. The need for a sounder theoretical basis for studies in urban history, followed by a more selective approach which focuses on those elements which will allow the theories to be tested, is one obvious gap to be filled by future research efforts.

One area where tentative steps have been made towards the development of theory is over the question of the 'modernity' of the nineteenth-century town. Though considered most explicitly in the essays on urban social structure, this is a topic which also has relevance for the other research areas represented in this volume. There have been a number of attempts to link nineteenth-century urban development to theories of social change (Timms, 1971; McGee, 1971; Shaw, 1977) but the degree to which the Victorian town was truly 'transitional' in its physical, economic and social characteristics between the 'pre-industrial' and 'modern' city has yet to be fully assessed. The question is of importance because in formulating theories relating to the internal structure of the Victorian city there are three possible positions which may be adopted. One is to take an evolutionary approach which places the transitional Victorian city on a continuum of change from pre-industrial to modern. A second is to view the Victorian town as a special case, a break with the past but still significantly different from the present. A third stance is to argue that the basic processes and the resulting patterns of urban form were broadly similar to those found in the twentieth-century city, so that theories of modern urban development will be applicable (Ward, 1975; Shaw, 1977; Carter and Wheatley, 1978c; Pooley, 1979b; Lewis, 1979). In order to make such assessments, massive problems of comparability of terminology and scales of analysis must be overcome, but none the less it is an important topic which has implications for most of the research themes outlined above.

A further problem which continually haunts any research on the internal structure of the nineteenth-century city is the introduction of a dynamic element into studies of urban history. All too often processes of historical change over time have been inferred from static cross-sections at decennial intervals. Although concentration on abundant census sources is understandable, by linking such data with other sources which have a shorter periodicity (but are not necessarily so complete) it should be possible to fill the gaps, at least partially. Studies of demographic change and residential mobility have gone some way towards providing a dynamic framework for the analysis of urban social change, but there are many other important processes causing change in the internal structure of the city and a more dynamic approach to all of these would be beneficial.

One problem which researchers from all disciplines share is that of providing an adequate explanation of the phenomena being studied. As already suggested, a number of alternative philosophies of investigation and associated explanation and interpretation are appropriate to the study of urban history. Whilst some problems may necessitate a traditional idiographic approach to historical research, others respond better to a positivistic philosophy and its associated methodology. Yet other problems may demand an idealistic approach which emphasises the role of individual actions and perceptions in shaping history, whilst a radical ideology which interprets the social problems of the nineteenth-century city within a Marxist framework is also applicable to the analysis of Victorian society (Guelke, 1974; Gregory, 1978; Foster, 1979). Clearly the philosophical and explanatory framework which is adopted will depend mainly on the personal conviction of an individual researcher, but some topics and sources do lend themselves more readily to one approach than another. Researchers from all disciplines should be sufficiently flexible to seek the explanatory framework most appropriate to the research in hand.

Lastly, although considerable emphasis has been placed on theory, philosophy and ideology in this introduction, one unifying theme which runs through all the approaches and topics discussed above is a practical concern about source availability and interpretation. Frequently, the most pertinent questions can only be partially answered because sources are simply not available, and interpretation must remain speculative because of doubts over the validity of the information. It is clear from the essays in this volume that a traditional historical concern with empirical research is not lost, and for further advances in urban history to be made much more empirical analysis is necessary. This research, however, must be married to a carefully-conceived philosophy of

24

historical investigation and a theoretical framework which directs empirical analysis towards the most pertinent questions.

Although there has been a tendency in this review to categorize approaches as being mainly geographical, historical or sociological, there are in fact many areas of interplay between the different disciplinary view-points. The contributions in this volume seek to facilitate this interaction. They also seek to highlight areas where further investigation is necessary, to suggest methods and techniques for empirical research and to generate theory which will allow some important questions about nineteenth-century urban structure to be answered within a comparative perspective.

References

Alexander, D. (1970) *Retailing in England during the industrial revolution*, Athlone, London

Alonso, W. (1964) *Location and land use: towards a general theory of land use*, Harvard University Press, Cambridge, Mass.

Anderson, G. (1976) *Victorian clerks*, Manchester University Press, Manchester

Anderson, M. (1971) *Family structure in nineteenth-century Lancashire*, Cambridge University Press, Cambridge

Anderson, M. (1976) 'Sociological history and the working-class family: Smelser revisited', *Social History*, 3, 317-34

Armstrong, W.A. (1974) *Stability and change in an English county town: a social study of York 1801-51*, Cambridge University Press, Cambridge

Ashton, T.S. (revised edition 1963) *Iron and steel in the industrial revolution*, Manchester University Press, Manchester

Ashworth, W. (1954) *The genesis of modern British town planning,* Routledge and Kegan Paul, London

Barker, T.C. & Harris, J.R. (1954) *A Merseyside town in the industrial revolution: St. Helens 1750-1900*, Liverpool University Press, Liverpool

Bater, J. (1976) *St Petersburg : industrialization and change*, Arnold, London

Benson, J. &. Neville, R.G. (eds) (1976) *Studies in the Yorkshire coal industry*, Manchester University Press, Manchester

Booth, C. (1889) *Life and labour of the people in London*, Macmillan, London

Briggs, A. & Gill, C. (1952) *The history of Birmingham*, Oxford University Press, Oxford

Briggs, A. (1963) *Victorian cities,* Odhams, London

Bowden, M. (1975) 'Growth of the central districts in large cities', in L.F. Schnore and E.E. Lampard (eds), *The new urban history: quantitative explorations by American historians*, Princeton University Press, Princeton, pp. 75-109

Brockington, C. (1965) *Public health in the nineteenth century*, Livingstone, Edinburgh

Burnett, J. (1974) *Useful toil*, Allen Lane, London

Burnett, J. (1978) *A social history of housing 1815-1970*, David and Charles, Newton Abbot

Bythell, D. (1969) *The handloom weavers*, Cambridge University Press, Cambridge

Cannadine, D. (1977) 'Victorian cities; how different?', *Social History,* 4, 457-82

Carter, H. (1965) *The towns of Wales: a study in urban geography*, Wales University Press, Cardiff

Carter, H. (1978) 'Towns and urban systems 1730-1900', in R.A. Dodgshon and R.A. Butlin (eds), *An historical geography of England and Wales*, Academic Press, London, pp. 367-400

Carter, H. & Wheatley, S. (1978a) 'Merthyr Tydfil in 1851: a study of spatial structure', *SSRC Project working paper 2*, Aberystwyth

Carter, H. & Wheatley, S. (1978b) 'Merthyr Tydfil in 1851: A study of spatial structure (2) - The grid square basis', *SSRC Project working paper 3*, Aberystwyth

Carter, H. & Wheatley, S. (1978c) 'Some aspects of the spatial structures of two Glamorgan towns in the nineteenth century', *Welsh History Review,* 9, 32-56

Carter, H. & Wheatley, S. (1979) 'Fixation lines and fringe belts, land uses and social areas: nineteenth-century change in the small town', *Transactions of the Institute of British Geographers,* N S 4, 214-238

Chadwick, E. (1842, 1965 edition) *Report on the sanitary conditions of the labouring population of Great Britain* (edited by M.W. Flinn), Edinburgh University Press, Edinburgh

Chalklin, C.W. (1974) *The provincial towns of Georgian England*, Arnold, London

Chaloner, W.H. (1950) *The social and economic development of Crewe, 1780-1923*, Manchester University Press, Manchester

Chapman, S.D. (1971) *The history of working-class housing*, David and Charles, Newton Abbot

Chapman, S.D. (1972) *The cotton industry in the industrial revolution*, Macmillan, London

Cherry, G.E. (1972) *Urban change and planning: a history of urban development in Britain since 1750*, Faulis, Henley

Cherry, G.E. (1979) 'The town planning movement and the late Victorian city', *Transactions of the Institute of British Geographers,* N S 4, 306-19

Chevalier, L. (1973, English edition) *Labouring classes and dangerous classes in Paris during the first half of the nineteenth century*, Routledge and Kegan Paul, London

Church, R.A. (1966) *Economic and social change in a midland town: Victorian Nottingham, 1815-1900*, Cass, London

Conzen, K.N. (1976) *Immigrant Milwaukee 1836-1860: accommodation and community in a frontier city*, Harvard University Press, Cambridge, Mass.

Conzen, M.R.G. (1960) 'Alnwick: a study in town plan analysis', *Transactions of the Institute of British Geographers*, no. 27, 1-122

Conzen, M.R.G. (1966) 'Historical townscapes in Britain: A problem in applied geography', in J.W. House (ed) *Northern Geographical Essays in honour of G.H.J. Daysh*, University of Newcastle Press, Newcastle, pp.56-78

Cowlard, K.A. (1979) 'The identification of social (class) areas and their place in nineteenth-century urban development', *Transactions of the Institute of British Geographers*, N S 4, 239-57

Creighton, C. (1894, 1965 edition) *A history of epidemics in Britain*, Cass, London

Crossick, G. (1976) 'The labour aristocracy and its values: a study of mid-Victorian Kentish London', *Victorian Studies*, 19, 301-28

Crossick, G. (ed) (1977) *The lower middle class in Britain*, Croom Helm, London

Crossick, G. (1978) *An artisan elite in Victorian society*, Croom Helm, London

Cunningham, C.M. (1970) 'The dock industry on Merseyside', in R. Lawton and C.M. Cunningham (eds), *Merseyside: social and economic studies*, Longman, London, pp.235-57

Daff, T. (1973) 'The establishment of iron-making in Scunthorpe 1858-77', *Bulletin of Economic Research*, 25, 104-21

Daunton, M.J. (1977) *Coal metropolis: Cardiff 1870-1914*, Leicester University Press, Leicester

Deane, P. and Cole, W.A. (1962) *British Economic Growth, 1688-1959*, Cambridge University Press, Cambridge

Dennis, R.J. (1977a) 'Distance and social interaction in a Victorian city', *Journal of Historical Geography*, 3, 237-250

Dennis, R.J. (1977b) 'Intercensal mobility in a Victorian city', *Transactions of the Institute of British Geographers*, N S 2, 349-63

Dennis, R.J. (ed) (1979) *The Victorian city. Transactions of the Institute of British Geographers*, N S 4, no. 2

Department of the Environment, (1977a) *Change and decay: final report of the Liverpool inner area study*, HMSO, London

Department of the Environment, (1977b) *Unequal city: final report of the Birmingham inner area study*, HMSO, London

Dewsnup, E.R. (1907) *The housing problem in England: its statistics, legislation and policy*, Manchester University Press, Manchester

Dillon, T. (1974) 'The Irish in Leeds 1851-1861', *Thoresby Society Publications Miscellany*, 16, 1-28

Dogan, M. and Rokkan, S. (1969) *Quantitative ecological analysis in the social sciences*, MIT Press, Cambridge, Mass.

Duncan, O.D., Cuzzort, R.P. and Duncan, B.A. (1961) *Statistical geography: Problems of analysing areal data*, The Free Press, Glencoe, Ill.

Dyos, H.J. (1961) *Victorian suburb: A study of the growth of Camberwell*, Leicester University Press, Leicester

Dyos, H.J. (1966) 'The growth of cities in the nineteenth century. A review of some recent writing', *Victorian Studies*, 9, 225-37

Dyos, H.J. (1967) 'The slums of Victorian London', *Victorian Studies*, 11, 5-40

Dyos, H.J. (ed) (1968) *The study of urban history*, Arnold, London

Dyos, H.J. and Reeder, D.A. (1973) 'Slums and suburbs', in H.J. Dyos, and M. Wolff (eds) *The Victorian city: images and realities*, Routledge and Kegan Paul, London, pp.359-86

Dyos, H.J. & Wolf, M. (1973) *The Victorian city: images and realities*, Routledge and Kegan Paul, London

Engels, F. (1845, 1969 edition) *The condition of the working class in England*, Panther, London

Farnie, D.A. (1979) *The English cotton industry and the world market, 1815-1896*, Clarendon Press, Oxford

Flinn, M.W. (1965) 'Introduction', in M.W. Flinn (ed), *The sanitary condition of the labouring population of Great Britain in 1842*, Edinburgh University Press, Edinburgh, pp.1-73

Floud, R. (1976) *The British machine-tool industry, 1850-1914*, Cambridge University Press, Cambridge

Forster, C.A. (1972) *Court housing in Kingston-upon-Hull*, University of Hull Press, Hull

Foster, J. (1974) *Class struggle and the industrial revolution: early industrial capitalism in three English towns*, Wiedenfeld and Nicolson, London

Foster, J. (1979) 'How imperial London preserved its slums', *International Journal of Urban and Regional Research*, 3, 93-114

Fox, R.C. (1979) 'The morphological, sociological and functional districts of Stirling, 1798-1881', *Transactions of the Institute of British Geographers*, N S 4, 153-167

Fraser, D. (1976) *Urban politics in Victorian England*, Leicester University Press, Leicester

Fraser, D. (1979) *Power and Authority in the Victorian City*, Blackwell, Oxford

Friedman, J. & Wolff, R. (1976) *The urban transition*, Arnold, London

Gatrell, V.A.C. (1977) 'Labour power and the size of firms in Lancashire cotton in the second quarter of the nineteenth century', *Economic History Review*, 30, 95-139

28

Gauldie, E. (1974) *Cruel habitations: A history of working-class housing*, Allen & Unwin, London

Geddes, P. (1915, 1949 edition) *Cities in evolution*, Williams and Norgate, London

Germani, G. (1969) 'Stages of modernization', *International Journal*, 24, 463-85

Gilbert, E.W. (1954) *Brighton: old ocean's bauble*, Methuen, London

Glaab, C.N. (1963) *The American city: a documentary history*, Dorsey Press, Homewood, Ill.

Goheen, P.G. (1970) *Victorian Toronto 1850-1900: pattern and process of growth*, Chicago University Press, Chicago

Gordon, G. (1979) 'The status areas of early to mid-Victorian Edinburgh', *Transactions of the Institute of British Geographers*, N S 4, 168-191

Gray, R.Q. (1976) *The labour aristocracy in Victorian Edinburgh*, Clarendon Press, Oxford

Gregory, D. (1978) *Ideology, science and human geography*, Hutchinson, London

Guelke, L. (1971) 'Problems of scientific explanation in geography', *Canadian Geographer*, 15, 38-53.

Guelke, L. (1974) 'An idealist alternative in human geography', *Annals of the Association of American Geographers*, 64, 193-202

Handlin, O. & Burchard, J. (eds) (1963) *The historian and the city*, MIT Press, Cambridge, Mass.

Harvey, D. (1973) *Social justice and the city*, Arnold, London

Hauser, P.M. & Schnore, L.F. (eds) (1965) *The study of urbanization*, Wiley, New York

Hennock, E.P. (1973) *Fit and proper persons: ideal and reality in nineteenth-century urban government*, Arnold, London

Hewitt, M. (1958) *Wives and mothers in Victorian industry*, Rockliff, London

Hill, E. (1975) *Victorian Lincoln*, Cambridge University Press, Cambridge

Hobsbawm, E.J. (1964) *Labouring men: studies in the history of labour*, Weidenfeld and Nicolson, London

Howard, E. (1898, 1965 edition) *Garden cities of tomorrow*, Faber, London

Hyde, F.E. (1971) *Liverpool and the Mersey: an economic history of a port 1700-1900*, David and Charles, Newton Abbot

Jeffreys, J.B. (1954) *Retail trading in Britain: 1850-1950*, Cambridge University Press, Cambridge

Katz, M.B. (1976) *The people of Hamilton, Canada West: family and class in a mid-nineteenth century city*, Harvard University Press, Cambridge, Mass.

Kay-Shuttleworth, J. (1832, 1970 ed.) *The moral and physical condition of the working classes employed in the cotton manufacture in Manchester*, Cass, London

Keat, R. and Urry, J. (1975) *Social theory as science*, Routledge and Kegan Paul, London

Kellett, J.R. (1969) *The impact of railways on Victorian cities*, Routledge and Kegan Paul, London

Kellett, J.R. (1978) 'Municipal socialism, enterprise and trading in the Victorian city', *Urban History Year-Book*, pp.36-45

King, L.J. (1976) 'Alternatives to a positive economic geography, *Annals of the Association of American geographers*, 66, 293-308

Kingsford, P.W. (1970) *Victorian railwaymen, 1830-1870*, Cass, London

Knights, P.R. (1971) *The plain people of Boston 1830-60*, Oxford University Press, New York

Lawson, R. (1973) *Brisbane in the 1890s: A study of an Australian urban society*, Queensland University Press, St Lucia, Queensland

Lawton, R. (1955) 'The population of Liverpool in the mid-nineteenth century', *Transactions of the Historic Society of Lancashire and Cheshire*, 107, 89-120

Lawton, R. (1958) Population movements in the West Midlands, 1841-1861, *Geography*, 43, 164-177

Lawton, R. (1978) *The census and social structure: an interpretative guide to nineteenth-century censuses for England and Wales*, Cass, London

Lawton, R. & Pooley, C.G. (1975a) 'Individual appraisals of nineteenth-century Liverpool', *Social geography of nineteenth-century Merseyside project, Working Paper* 3, Liverpool

Lawton, R. & Pooley, C.G. (1975b) 'The urban dimensions of nineteenth-century Liverpool', *Social geography of nineteenth-century Merseyside project, Working Paper* 4, Liverpool

Lawton, R. & Pooley, C.G. (1976) *The social geography of Merseyside in the nineteenth century*, final report to the SSRC, Liverpool

Lees, L.H. (1969) 'Patterns of lower-class life: Irish slum communities in nineteenth-century London', in S. Thernstrom and R. Sennett (eds), *Nineteenth-century cities*, Yale University Press, New Haven, pp.359-85

Lees, L.H. (1973) 'Metropolitan types: London and Paris compared', in H.J. Dyos and M. Wolff (eds) *The Victorian city: images and realities*, Routledge and Kegan Paul, London, pp.413-28

Lees, L.H. (1976) 'Mid-Victorian immigration and the Irish family economy', *Victorian Studies*, 20, 25-44

Lees, L.H. (1979) *Exiles of Erin*, Manchester University Press, Manchester

Lewis, C.R. (1979) 'A stage in the development of the industrial town: a case study of Cardiff, 1845-75', *Transactions of the Institute of British Geographers*, N S 4, 129-52

Lewis, R.A. (1952) *Edwin Chadwick and the Public Health Movement, 1832-52*, Longman, London.

Lipman, V.D. (1954) *Social history of the Jews in England*, Watts, London

30

Lobban, R.D. (1971) 'The Irish community in Greenock in the nineteenth-century', *Irish Geography*, 6, 270-81.

Lovell, J. (1969) *Stevedores and Dockers: a study of trade unionism in the port of London, 1870-1914*, Macmillan, London

Macfarlane, A. (1977a) 'History, anthropology and the study of communities', *Social History*, 4, 631-52

Macfarlane, A. (1977b), *Reconstructing historical communities*, Cambridge University Press, Cambridge

McGee, T.G. (1971) *The urbanization process in the Third World*, Bell, London

Mayhew, H. (1851, 1967 edition) *London labour and the London poor*, Cass, London

Morris, R.J. (1979) *Class and class consciousness in the Industrial Revolution, 1780-1850*, Macmillan, London

Mortimore, M. (1969) 'Land ownership and urban growth in Bradford and environs in the West Riding Conurbation 1850-1950', *Transactions of the Institute of British Geographers*, 46, 105-20

Musson, A.E. and Robinson, W.G. (1959) 'The early growth of steam power', *Economic History Review*, 11, 418-39

Neale, R.S. (1972) *Class and ideology in the nineteenth century*, Routledge and Kegan Paul, London

Ogden, P.E. (1977) 'Foreigners in Paris: residential segregation in the nineteenth and twentieth centuries', *Department of Geography, Queen Mary College, Occasional paper*, 11, London

Olsen, D.J. (1976) *The growth of Victorian London*, Batsford, London

Pahl, R.E. (1970) *Patterns of urban life*, Longman, London

Parry-Lewis, J. (1965) *Building cycles and Britain's growth*, Manchester University Press, Manchester

Patmore, J.A. (1962) 'A Navvy gang of 1851', *Journal of Transport History*, 5,

Perkin, H. (1969) *The origins of modern English society*, Routledge and Kegan Paul, London

Perkin, H. (1970) *The age of the railway*, Routledge and Kegan Paul, London

Pollard, S. (1959) *A history of labour in Sheffield*, Liverpool University Press, Liverpool

Pooley, C.G. (1977) 'The residential segregation of migrant communities in mid-Victorian Liverpool', *Transactions of the Institute of British Geographers*, N S 2, 364-82

Pooley, C.G. (1979a) 'Residential mobility in the Victorian city', *Transactions of the Institute of British Geographers*, N S 4, 258-277

Pooley, C.G. (1979b) 'Residential differentiation in the nineteenth-century city', in B.C. Burnham and J. Kingsbury (eds) *Space, Hierarchy and Society*, B.A.R. International Series 59, Oxford, pp.161-86

Prince, H.C. (1964) 'North-west London 1814-1863' and 'North-west London 1864-1914', in J.T. Coppock and

H.C. Prince (eds) *Greater London*, Faber, London, pp.80-141

Pritchard, R.M. (1976) *Housing and the spatial structure of the city*, Cambridge University Press, Cambridge

Rathbone, E.F. (1904) *Report of an enquiry into the conditions of dock labour at Liverpool Docks*, Liverpool

Richardson, C. (1968) 'Irish settlement in mid-nineteenth-century Bradford', *Yorkshire Bulletin of Economic and Social Research*, 20, 40-57

Reissman, L. (1964) *The urban process: cities in industrial societies*, The Free Press, Glencoe, Ill.

Roberts, E. (1976) 'Working-class Barrow and Lancaster, 1890-1930', *Centre for North-West Regional Studies, Occasional Paper* 2, Lancaster

Roberts, R. (1971) *The classic slum, Salford life in the first quarter of the century*, Manchester University Press, Manchester

Robinson, W.S. (1950) 'Ecological correlations and the behaviour of individuals', *American Sociological Review*, 15, 351-7

Robson, B.T. (1973) *Urban growth: an approach*, Methuen, London

Rosen, G. (1971) 'Social variables and health in an urban environment: The case of the Victorian city', *Clio Medica*, 8, 1-17

Rosen, G. (1973) 'Disease, debility and death', in H.J. Dyos and M. Wolff (eds), *The Victorian city: images and realities*, Routledge and Kegan Paul, London, pp. 625-67

Rostow, W.W. (1960) *The stages of economic growth*, Cambridge University Press, Cambridge

Scola, R. (1975) 'Foodmarkets and shops in Manchester, 1770-1870', *Journal of Historical Geography*, 1, 153-68

Sharpless, J.B. (1976) 'The economic structure of port cities in the mid-nineteenth-century: Boston and Liverpool 1840-1860', *Journal of Historical Geography*, 2, 131-45

Shaw, G. & Wild, M.T. (1979) 'Retail patterns in the Victorian city', *Transactions of the Institute of British Geographers*, N S 4, 278-91

Shaw, M. (1977) 'The ecology of social change: Wolverhampton 1851-71', *Transactions of the Institute of British Geographers*, N S 2, 332-48

Shaw, M. (1979) 'Reconciling social and physical space: Wolverhampton 1871', *Transactions of the Institute of British Geographers*, N S 4, 192-213

Simpson, M.A. & Lloyd, J.H. (eds) (1977) *Middle-class housing in Britain*, David and Charles, Newton Abbot

Smelser, N.J. (1959) *Social change in the industrial revolution*, Routledge and Kegan Paul, London

Smith, F.B. (1979) *The people's health, 1830-1910*, Croom Helm, London

Smith, N.A.F. (1975) 'Nineteenth-century civil engineering', *History of Science*, 13, 105-13

Stedman-Jones, G. (1971) *Outcast London: A study in the*

relationship between classes in Victorian London, Oxford University Press, Oxford

Straw, F.I. (1967) 'An analysis of the town plan of Nottingham: a study in historical geography', Unpub. M.A. thesis, University of Nottingham

Supple, B. (ed) (1977) *Essays in British business history*, Clarendon Press, Oxford

Sutcliffe, A. (1970) *The autumn of central Paris*, Arnold, London

Sutcliffe, A. (ed) (1974) *Multi-storey living: the British working-class experience*, Croom Helm, London

Sutcliffe, A. (1981) *Towards the planned city: Germany, Britain, the United States and France, 1780-1914*, Blackwell, Oxford

Tansey, P.A. (1973) 'Residential patterns in the nineteenth-century city: Kingston-upon-Hull 1851', Unpub. Ph.D. thesis, University of Hull

Taplin, E. (1974) *Liverpool dockers and seamen*, Hull University Press, Hull

Tarn, J.N. (1971) *Working-class housing in nineteenth-century Britain*, Lund Humphries, London

Tarn, J.N. (1973) *Five-per-cent philanthropy*, Cambridge University Press, Cambridge

Taylor, I.C. (1976) 'Black spot on the Mersey', Unpub. Ph D thesis, University of Liverpool

Thernstrom, S.A. (1964) *Poverty and progress: social mobility in a nineteenth-century city*, Harvard University Press, Cambridge, Mass.

Thernstrom, S.A. (1973) *The other Bostonians. Poverty and progress in the American metropolis: 1880-1970*, Harvard University Press, Cambridge, Mass.

Thomas, B. (1972) *Migration and urban development*, Methuen, London

Thompson, E.P. (1963) *The making of the English working class*, Gollancz, London

Thompson, F.M.L. (1974) *Hampstead: building a borough, 1650-1964*, Routledge and Kegan Paul, London

Thompson, P. (1973) 'Voices from within', in H.J. Dyos and M. Wolff (eds), *The Victorian city: images and realities*, Routledge and Kegan Paul, London, pp.59-80

Timms, D.W.G. (1971) *The urban mosaic*, Cambridge University Press, Cambridge

Urban History Year-Book (1974-), University Press, Leicester

Vance, J.E. (1967) 'Housing the worker: determinative and contingent ties in nineteenth-century Birmingham', *Economic geography*, 43, 95-127

Vaughan, R. (1843, 1972 edition) *The age of great cities, or modern society viewed in relation to intelligence, morals and religion*, Irish University Press, Shannon

Walton, J.K. (1974) 'The social development of Blackpool 1788-1914', Unpub. Ph.D.thesis, University of Lancaster

Ward, D. (1962) 'The pre-urban cadastre and the urban pattern of Leeds', *Annals of the Association of American Geographers*, 52, 150-66

Ward, D. (1964) 'A comparative historical geography of streetcar suburbs, Boston ... and Leeds ... 1850-1920', *Annals of the Association of American Geographers*, 54, 477-89

Ward, D. (1966) 'The industrial revolution and the emergence of Boston's central business district', *Economic Geography*, 42, 152-71

Ward, D. (1971) *Cities and immigrants*, Oxford University Press, New York

Ward, D. (1975) 'Victorian cities. How modern?', *Journal of Historical Geography*, 1, 135-51

Ward, D. (1980) 'Environs and neighbours in the 'Two Nations': residential differentiation in mid-nineteenth century Leeds', *Journal of Historical Geography*, 6, 133-62

Warnes, A.M. (1972) 'Residential patterns in an emerging industrial town', in B.D. Clark and M.B. Gleave (eds), 'Social patterns in cities', *Institute of British Geographers, special publication* no. 5, 169-89

Weber, A.F. (1899, 1963 edition) *The growth of cities in the nineteenth century*, Cornell University Press, New York

Werly, J. (1973) 'The Irish in Manchester', *Irish historical studies*, 18, 345-58

Whitehand, J.W.R. (1967) 'Fringe belts: a neglected aspect of urban geography', *Transactions of the Institute of British geographers*, no. 41, 223-33

Whitehand, J.W.R. (1972) 'Building cycles and the spatial pattern of urban growth', *Transactions of the Institute of British geographers*, 56, 39-55

Whitehand, J.W.R. (1974) 'The changing nature of the urban fringe' in J.H. Johnson (ed), *Suburban growth: geographical processes at the edge of the Western city*, Wiley, London, pp.31-52

Whitehand, J.W.R. (1975) 'Building activity and intensity of development at the urban fringe: the case of a London suburb in the nineteenth century', *Journal of historical geography*, 1, 211-24

Whitehand, J.W.R. (1977) 'The basis for an historico-geographical theory of urban form', *Transactions of the Institute of British geographers*, N S 2, 400-16

Whitehand, J.W.R. (1978) 'Long-term changes in the form of the city centre: the case of redevelopment', *Geografisca Annaler*, 60 B, 79-96

Whitehand, J.W.R. and Patten, J. (eds) (1977) *Change in the town, Transactions of the Institute of British Geographers* N S 2, no. 3

Wild, T. (ed) (1980) *Register of research in Historical Geography*, Geo Abstracts, Norwich

Wild, M.T. & Shaw, G. (1974) 'Locational behaviour of urban retailing during the nineteenth century: the example of Kingston-upon-Hull', *Transactions of the Institute of*

34

British geographers, 61, 81-100

Williams, M. (1974) *The making of the South Australian landscape*, Academic Press, London

Williams, B. (1976) *The making of Manchester Jewry, 1740-1875*, Manchester University Press, Manchester

Wohl, A. (1977) *The eternal slum: housing and social policy in Victorian London*, Arnold, London

Woods, R. (1978) 'Mortality and sanitary conditions in the "best governed city in the world" - Birmingham, 1870-1910', *Journal of historical geography*, 4, 35-56

Woodward, J. & Richards, D. (eds) (1977) *Health and popular medicine in nineteenth-century England: essays in the social history of medicine*, Croom Helm, London

Wrigley, E.A. (ed) (1966) *An introduction to English historical demography*, Weidenfeld and Nicholson, London

Wrigley, E.A. (ed) (1972) *Nineteenth-century society: essays in the use of quantitative methods for the study of social data*, Cambridge University Press, Cambridge

Yelling, J.A. (1981) 'The selection of sites for slum clearance in London, 1875-1888', *Journal of Historical Geography*, 7, 155-65

PART TWO

HOUSING AND THE NINETEENTH-CENTURY
URBAN ENVIRONMENT

The three studies of the physical structure of the
nineteenth-century city presented in this section
concentrate on housing and the urban environment,
but each approaches the topic from a quite
different point-of-view.

Richard Rodger adopts an explicitly economic
framework for his study of the urban land market,
arguing that macro-economic models continue to have
considerable relevance for our understanding of the
nineteenth-century land market despite the increasing number of factors which interfered with and
distorted the classical operation of the land market
in the second half of the century. Although his
interpretation did not go unchallenged in discussion
during the seminar, his case study of the Scottish
urban land market provides a valuable example of
the macro-economic study of this topic.

In contrast, Peter Aspinall concerns himself
with the internal organization of the nineteenth-
century building industry. Through a careful case-
study based on Sheffield, he demonstrates how the
changing structure of the industry affected the
supply of housing in different parts of the market
and hence the eventual form of nineteenth-century
towns. This aspect of the supply of housing -
focusing on the building industry itself rather than
the economics of land development - is an interesting approach which deserves further exploration.

In the third essay Anthony Sutcliffe is concerned with much wider aspects of the nineteenth-
century urban environment. He also emphasizes the
way in which an increasing number of factors began
to influence all aspects of the urban environment in
the second half of the century, but concentrates
particularly on the course of housing and health
reforms. He suggests a framework in which increasing municipal intervention may be viewed as a direct

response to a malfunctioning of the urban environment and increasing disatisfaction with urban living conditions. The questions of why, when and where municipal authorities chose to intervene in urban reform are discussed, and the links between nineteenth-century urban reform and the development of the modern planning process are explored.

Many other aspects of the urban environment could have been examined. There is, for instance, relatively little here on urban disease, on the micro-scale development of building plots, or on the operation of the nineteenth-century housing market. Nevertheless, the three essays which are presented open up a wide range of topics, and each has implications for the future direction of research on housing and the urban environment. Clearly the built environment offers an essential context within which studies of the social and economic characteristics of the urban population must be placed. However, the physical form of the city provided no passive backcloth. It was constantly evolving; and the morphological and environmental characteristics of different areas profoundly affected the social and economic composition of different parts of the city.

Chapter 2

RENTS AND GROUND RENTS : HOUSING AND THE
LAND MARKET IN NINETEENTH-CENTURY BRITAIN

Richard Rodger

Introduction

The role of the land market in the provision of
urban living space is a neglected and yet important
determinant of housing type and quality, providing
the initial point of intersection between the
demand for and supply of urban space. It is,
however, only one factor amongst many that could be
considered. Demand for urban housing was affected
by the pace of urban demographic change (both in
terms of absolute growth and structural change),
while the growth of real incomes and their susceptibility to short period fluctuations must also be
considered. From a supply viewpoint, the responses
of builders to changes in variables such as the
availability of credit, the extent of market satura-
tion, the availability of transport facilities and
the structure of the building industry and its
firms, also affected the quality and quantity of
housing ultimately produced. Of the components of
weekly rental payments - repairs, maintenance,
insurance, land and development costs, interest,
collection and administration charges - the element
which covered the cost of the land (ground rents)
represented only one-fifth or one-sixth of the
total weekly rental payments.[1] Thus the influence
of the land element must not be overstated in the
complex array of factors which determined
Victorian housebuilding decisions. Emphasis in
this essay is placed on the role of the land market
as an influence on the fabric and internal structure
of Victorian cities,but this in no way attempts to
deny the importance of other factors, or discredit
the powerful analyses of effective demand fluctuations and the lagged adjustment process of an
industry with a peculiarly durable product
(Matthews, 1959; Saul, 1962).

Research into housing form and development in
Britain has tended to concentrate on the idiosyn-
cracies of a particular urban area, eschewing the
determinism of a general model of market equilibrium
for land.   Thus the concept of land as a factor
market, in which the forces of demand and supply
determine an equilibrium price which will clear the
market and thus dictate land usage, is normally
neglected as attention is directed at the composition of demand and the nature of supply.   The
constituent items of the housebuilding equation
such as population change,trade-cycle fluctuations,
institutional rigidities and local government
decisions, and the attitudes of landowners to
estate developments have been individually assessed,
but the final solution of the equation, the price
of land, has been less than fully considered.[2]
Whilst the relevance of a pricing mechanism might
be acknowledged, most studies have emphasized the
peculiarities of topography, transport, class
composition and the industrial base in a particular
city, as well as aesthetic considerations of
architects and landowners.   Thus an individualistic
approach to urban development has emerged stressing
the evolutionary basis of expansion and the
importance of small-scale analysis.   This concentration on the individuality of estate development in
British studies of urban development is conspicu-
ously different to most North American work which
has rejected as moribund the idiosyncratic emphasis
on this side of the Atlantic, and has embraced a
more general framework for urban analysis (Lowry,
1967).
     Whatever the methodological merits of case
studies versus general models, it is the contention
of this essay that factors such as estate development strategies of individual landowners, the role
of transport developments, the burgeoning local
authority requirements for land, and other
influences on the urban development of particular
areas can be regarded as special cases - admittedly
numerous - which amend rather than overthrow the
fundamental dynamics of an urban land market.   The
price mechanism may be relegated to a position of
secondary importance in any given outcome of land
usage, but that should not obscure its underlying
influence in explanations of residential differentiation, urban zoning and related issues.   It is
thus suggested that the peculiarities and apparent
inconsistencies in a land market model of urban
space, represent no more than imperfect solutions
to the theoretical workings of the land market.

The land-market model

The most robust of the land market models relies on
a form of indifference curve analysis in which con-
sumers' preferences (private and corporate) for
access to the city centre are mapped against their
respective ability to pay for central sites (Segal,
1977). Value and utility are therefore ascribed
to various locations within a city and this is
measured in terms of the willingness and ability of
different users to pay for each type of site. The
intrinsic value of land was presented in the
following terms by Ricardo: '...its total supply
being inelastic the land will always work for whatever is given to it under competition'. Von
Thünen further developed the relationship between
value and rents by introducing the concept of rent
differentials, even on uniformly fertile agricultural land, based on relative distance from the
market (Dickinson, 1969). This consequently
introduced different patterns of crop as land usage
varied according to the location of any particular
site in relation to the marketing centre.
Extension of this model to the urban case is
straightforward: 'If we investigate the reasons why
site rent increases towards the center of the city
we will find it is the labour saving, the greater
saving, the greater convenience and the reduction
of the loss of time in connection with the pursuit
of business' (Ely and Wehrwein, 1940, p.444). A
simplification of this position was offered by Hurd
(1903, p.13), for, 'since value depends on economic
rent and rent on location and location on nearness,
we may eliminate the intermediate steps and say
that value depends on nearness.' Rent,therefore,
which can be capitalized to give an equivalent land
value and purchase price, is a charge imposed by
the landowner for the intrinsic value of his site,
namely, 'the saving in transport costs which the
use of his site makes possible' (Alonso, 1964, p.6).
Where no such savings accrue the built-up area ends,
and land reverts to agricultural usage.

In general terms, accessibility to central
sites creates a graduated pattern of land values
declining from city centre to the urban fringe, and
the development of any particular site represents
the user's reconciliation of accessibility requirements and their rent-paying capacity. This relationship could exist regardless of whether they are
seeking to maximize utility, as postulated by
Alonso (1960), or to minimize costs of transport, as

suggested by Kain (1962), Wingo (1961) and others.
In general, landowners will attempt to maximize
their land values, and this will interact with
buyers' evaluation of the convenience of sites in
relation to the city centre, and their respective
abilities to pay for such sites to determine land
use.   Hence retailers, for whom accessibility to
the entire urban population is at a premium, are
prepared to pay very high rents as their potential
turnover suffers appreciably with distance from the
city centre.   Thus the 'trade-off', or the gradient
of the substitution between rental capacity and
distance is extremely steep (Figure 2.1). Office,
commercial and professional uses, whilst preferring
central locations, are not so prejudiced if their
activities are conducted in properties adjacent to
the High Street.   This is reflected in the less
steep gradient of rental ability in relation to
accessibility for these users in Figure 2.1.   The
Victorian labour market, typified by casual daily
employment, long hours and the absence of canteen
facilities at work, placed a premium on accessibility to the central business or industrial district
for the majority of the working population.   High
rise and high density dwellings on the edge of the
commercial area were viable solutions to this
requirement, and the sub-division of high rentals
amongst many households offered a means of achieving
this on sites which were expensive because of the
proximity to the central urban core.   Multiple
housing units and high-density accommodation thus
possessed steeper rent-distance gradients than
single-family dwellings whose occupants, because of
different employment terms, social needs and
abilities to pay transport costs, could utilize a
more distant location from the city centre. Furthermore, the single-family household was less able to
pay high rentals for central locations as these
were not divisible amongst several users as in the
case of tenement dwellers.   Finally, agriculture
would also benefit from access to the market, but
as this land use is least capable of matching the
rentals payable by urban users the gradient is
least steep (Figure 2.1) (Alonso,   1960).   When
the relative capabilities to pay rents are superimposed in Figure 2.2, then the overall urban land-
use pattern is determined.   Landowners, wishing to
maximize their receipts, opt for the rentals
represented by the outermost points on the combined
rental curve (OABCDEI).   The actual division of the
land between users is determined by the competitive
bidding process, and by the collective and

42

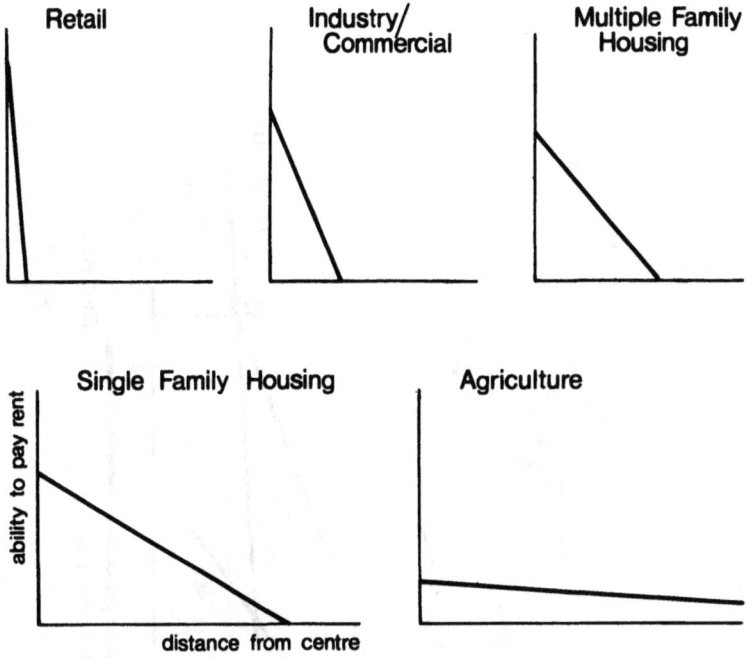

Figure 2.1   Individual land-users' rent-distance substitutions

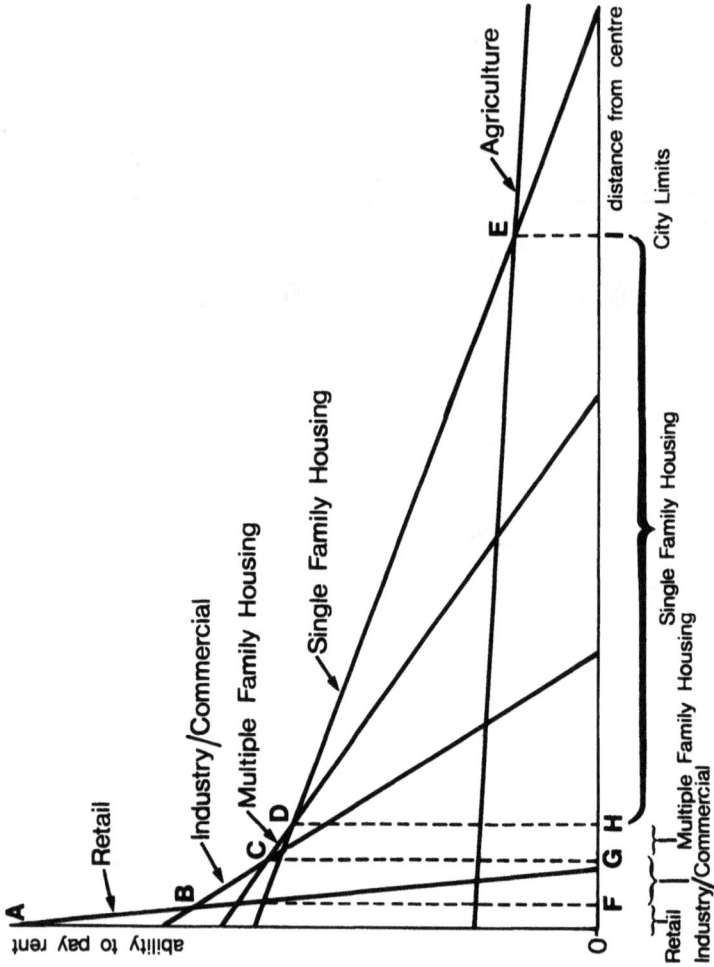

Figure 2.2 Composite rent-distance substitutions: various land uses superimposed

individual reconciliations of distance and rent.
Hence just beyond OF in Figure 2.2 retailers find
proceeds decline so rapidly that they neither want
to, nor can pay for such sites, whilst even when
divided amongst a large number of high-density
dwellings the rental is still beyond the residents'
ability to pay, thereby leaving the land between OF
and OG to business uses. As points F, G, H and I
are successively approached there will be keener
competition for land as the marginal advantage or
disadvantage becomes less obvious for each land use.
Nonetheless, the theoretical outcome should be a
series of discrete land uses although in practice
some overlap tends to result at the various rental
boundaries. Furthermore, this land-market model
applies to all towns, although its actual operation
and the land uses determined will vary according to
the composition and strength of competing interests.
     The process of functional specialization in
land use is, however, an evolutionary one. Unless
there is a substantial existing supply of central
land, or an infinite capacity for multi-storey
building, then urban expansion must, by definition,
produce sites which are successively more remote
from the city centre. Hence rents for central
locations will move upwards as their accessibility
commands a further premium. According to Burgess
(1925, p.49) 'the tendency of each inner zone to
extend its area by invasion of the next outer zone'
typifies this organic scheme of concentrically
arranged land use. Although Burgess' evolutionary
model of land-use rings has been vigorously attacked
(Davie, 1961), it has been successively refined and
adapted, as for example by Hoyt (1939) in his
presentation of sectoral intrusions into the concentric pattern, and by Harris and Ullman (1945)
with their identification of a multi-nuclei model.
As such the fundamental relationship between
accessibility and rentals appears dented but
durable.
     Indeed, Taylor (1949) has provided a further
dynamic dimension to the operations of the land
market by identifying various stages of urban
development which he termed infantile, juvenile,
adolescent, early mature and mature towns. This
has contributed a degree of reality and flexibility
which the model previously lacked in its explanation of functional specialization. Population
growth and urban expansion introduce complexities
into manufacturing and residential land requirements which are most easily accommodated by spatial
concentration of land uses. Only beyond some

notional threshold urban size does such specialization in land use operate. The issue of accessibility, and thus site rent, rarely arises in
villages or even small towns, where transport costs
and savings are small. It is, however, difficult
to be precise about the absolute size of town at
which the land market model is activated, though
early modern British towns do exhibit some features
of zoning (Sjoberg, 1976).

The internal structure of nineteenth-century
British cities possesses many common denominators,
based to some degree on functional specialization in
land use. Although founded on American experience
and despite a recent contrary view (Cannadine, 1977),
the conclusion of Ratcliffe (1949, p.369) is still
instructive. He argues that 'the same basic
tendencies appear in all cities in spite of minor
differences resulting from variation in topography,
size and maturity. The underlying pattern is
apparent even though it is often modified by
irrational real estate developments or special
physical conditions.' Nineteenth-century British
housing must therefore be seen in the light of
evolutionary pressure on scarce central space, and
in the wider context of competing land usage. The
ultimate internal physical structure reflects the
relative pecking order of ability to pay rent for
the privilege of accessibility to the urban centre
and the advantages that that confers.

Complications to the land-market model

Greater realism and historical relevance can be
introduced into the schema already advanced by
incorporating changes in population, real income
and technological development in urban transport.
Improvements in real income or in the urban
transport network affect the land market in that
commuting may be both easier and cheaper, and consequently the quality of housing, or the affordable
plot size, can be increased. These trends were
evident in Victorian Britain, and suburban developments in many towns reflect this influence. Put
in the context of the land market, the effect of
increased effective demand, for whatever reason, is
to flatten the gradient of the rent - distance
curve (Wheaton, 1974). The disutility of distance
is diminished and land prices at the urban fringe
are increased. Thus a centrifugal force is
introduced as transport or real income improvements
reduce the relative advantages of central location

and the margin of the built-up area is extended.

Population growth, by contrast, offers a reversal of the stimulus. An increase in population (whether from immigration, natural growth or structural changes in the pattern of household formation) represents additional demand for residential space. Thus the existing population must make higher rental bids for land in order to retain their current space and location. Increased household formation normally competes for areas of greatest accessibility to existing employment opportunities and thus higher rental curves operate in all locations, though notably in the central area. The effect of population increases is thus to raise rentals in particular areas with ripple or echo effects communicated to other land users (Beckman, 1969). Population growth, by introducing friction into the market and into established land-use patterns, also tends to force overlapping rental bids at the boundaries of different land uses. This is most noticeable in areas of high residential status which are gradually eroded under pressure of population growth, as the margin of multiple-family and high-density dwelling advances.

The nineteenth-century town thus experienced conflicting forces. Put simply:'An increase in the demand for urban land ... tends to raise land prices. On the other hand, decentralization - which increases the supply of usable land - tends to lower prices' (Hallett, 1977, p.105). The historical reconciliation of these opposing demographic, income and transport influences therefore depended on local circumstances and the extent to which one influence held sway at any particular time.

Certainly transport improvements tended to be sporadic. This discontinuity led Clarke (1965, p.139) to argue that: 'It was the arrival of the electrically driven tramcar in the 1890s which began to chisel apart the compact Victorian city, reducing population densities and at the same time reducing relative land values.' Taken together with intensifying demands for land-taxation reforms, this would seem to suggest that in the battle between opposing forces operating on land prices the role of demand for urban space was more influential.

Sophistications in the operations of the land market should also include the concept of changing factor intensities, both spatially and historically. Stigler (1966, p.54) expresses the issue of different factor combinations succinctly. Implications for land usage and residential differentiation are based on the proposition that: 'Since land

becomes cheaper the more distant it is from the
center, but building declines much less, if at all,
as one moves out, land will be used in larger
quantities relative to buildings.   Hence the
density of population will also decrease as the
distance from the city center increases.' This
spatial variation in factor inputs usefully
supplements the 'ability to pay rent' approach of
Alonso (1960) and others in attempting to explain
land uses.   Whitehand (1977, p.407) uses a similar
approach, but relates it to a time dimension,
particularly to the ebb and flow of the building
cycle.   'Housebuilders', he argues, 'are likely to
effect greater economies in the use of land when
this is expensive relative to building costs (as is
generally the case during a building boom) by
reducing plot sizes.   During building slumps the
opposite tendency is to be expected since falls in
land values tend on the whole not to be associated
with proportionate falls in building costs.'   Thus
both within and between categories of building, the
flexibility of land costs relative to other
building costs achieves a differential impact
according to whether the proposed buildings are
land intensive or land extensive.   For example,
institutional building which is land extensive,
tends to be undertaken on a counter-cyclical basis
to housebuilding.

It should be noted that while this aspect of
a cyclical and spatial functioning of the land
market is instructive, it is an extension of the
central propositions previously advanced.   Thus
the ideas of Whitehand can be internalized within
the general land-market model.   Institutional
building is undertaken during building depressions
because, utilizing the relative factor-cost
argument, its extensive input, land, is least
expensive in this cyclical phase.   But it is also
because institutional ability to pay (rent, or its
capitalized version, purchase price) for central,
high-value sites is limited, and these users must
necessarily resort to more distant locations.

Emphasis has been placed on a factor market in
which the saleable item, land, has an intrinsic
value, location, with respect to some focal point.
Accessibility in the form of transport savings
commands a premium in the form of higher rents
which can be progressively substituted, less rent
generally being associated with less accessibility.
Population changes, technical developments in
transport, alterations in effective demand and
fluctuations in building activity will all alter

the utility, value and thus the cost of sites over
time.   Infinite flexibility does not apply
because of the commitment to existing uses on
previously developed sites.   Furthermore, values
and preferences may change appreciably over time
and between countries.   For instance, the
desirability of central flats has fluctuated in
Georgian Edinburgh (Youngson, 1966) and in certain
quarters of late-nineteenth century Paris (Sutcliffe,
1970), while cultural values regarding gardens have
also been subject to variations (Tarn, 1973).
Thus the perfect operation, or the universal
application of a pricing mechanism in the determination of land uses should not be too strongly
pressed.   Partial information, the absence of a
central exchange and the inflexibility of existing
land use introduced distortions into the land
market.   Only rarely were auctions conducted, the
price was often a closely guarded secret and
simultaneous availability of all land never existed.
Not surprisingly, the eventual urban configuration
was highly individualistic, but this by no means
destroys the underlying relevance of the market
mechanism.

The influence of landowners on the land market

Studies of imperfections in the land market usually
concentrate on the haemorrhaging of supplies of
suitable land for working-class housing.   Indeed
this was a major preoccupation of Victorians
seeking to explain the seemingly intractable
problem of housing the poorest groups in society.
It would seem an over-simplification, however, to
argue as one contemporary did, that 'the principal
cause of high rents (was) the land monopoly.' [3]
Nevertheless, the ability of the landowner to
influence the supply of land and its price was
sometimes considerable.   In Liverpool, Edinburgh,
Birmingham, Cardiff and London there were several
examples of developers withholding building land
with a view to increasing its scarcity value.
According to one expert, land speculators 'have
often made several hundred per cent profit at the
expense of the health and well-being of the nation.
If the poorest classes are to be housed satisfactorily at reasonable rents these excessive
profits will have to cease.' [4]   The grip of landowners could also be decisive.   In Clydebank for

instance, where 'most of the land within the burgh is held by six separate owners' the opportunity for "land sweating" was clearly much greater than in towns where ownership patterns were more fragmented and in Glasgow in 1903, 'large syndicates (were) eagerly buying up the land in districts opened up or likely to be opened up by the tramway system.'[5] Such practices, yielding sizeable gains, merely endorsed the mechanics of the land market. As greater utility was attached to specific sites because of transport improvements and the enhanced accessibility to central areas, rental levels could be raised. Landowners thus capitalized the higher rental capacities of their sites and achieved substantial windfalls, which were the monetary expression of the transport extensions. Occasionally, and perhaps more significantly, the same landowners through patronage and political leverage were able to influence the projected routes for urban transport developments and could accordingly determine precisely who the beneficiaries would be.[6]

Interruption of the supply of building land was a problem which was further exacerbated by the prevailing system of local taxation. Rates were levied according to the existing land use and the rentals associated with this. Hence taxation effectively reinforced the existing land-use zoning by differentiating central, suburban and agricultural uses within the city and taxing them accordingly.[7] In Glasgow, in the assessment year of 1912-13, the average built-up area valuation was £522 per acre; vacant land suitable for building purposes was valued at £5.9s.8d. per acre and 'agricultural' land within the burgh boundary at £2.1s.6d. per acre.[8] Municipal rates paid per acre were respectively, £99; 5s.6d. and 1s.7d. The incentive to withhold land was often irresistible, for the landowner paid only nominal rates on land which, once developed, brought him a substantial capital gain.

The partial and controlled release of land was thus encouraged by valuation procedures which not only failed to penalize the practice of withholding land but actively encouraged it. Such a stranglehold on building land raised the overall value of existing sites as well as those available for development, and acted as a force for increased rentals and purchase prices throughout the spectrum of urban locations and land uses. It also, not surprisingly, inflamed contemporary feeling and spawned reformist movements advocating betterment taxation.[9]

50

To place landowners in the context of a malfunctioning land market, where inflated land prices
produced attendant problems of high density, low
amenity dwellings for working-class groups via the
level of ground rents, is both to oversimplify the
relationship and to overlook the fact that they
were abiding by the framework of an unrestrained
land market.   The utility of a site was expressed
monetarily in its rental capacity and landowners
were thus merely receiving the market evaluation of
the site value.   In addition, the term 'malfunctioning' must be used guardedly.   In mid-Victorian
Britain the laissez-faire posture of local and
central government (Wohl, 1977) entailed only
marginal concern about the distributional effects
of the operation of the land market.   For various
reasons - including arguments of national
efficiency, and ideas of equity and social welfare
(Hay, 1975) - a different stance was increasingly
adopted towards working-class housing and an
uncontrolled land market as the nineteenth century
progressed.   The nature and quality of low-income
housing was an omni-present topic, and the distributional implications of the pricing mechanism in
the land market were successively attacked in
various statutes.   Institutional constraints,
including building byelaws, local improvement
schemes and town-planning projects, local-authority
building, betterment and site-value taxation
reforms, were limitations introduced to mitigate
the worst environmental consequences of high-
density housing which were partially attributable
to the functioning of the land market.   Towards
the end of the nineteenth century changing social
values thus began to interrupt the pricing
mechanism in the land market which still retained
its fundamental operation, although it did not
remain completely unchecked. [10]
     It was, however, precisely these regulatory
attempts by local and central government which
created many of the difficulties and anomalies in
the land market, and which achieved little of any
consequence in the field of working-class housing
in the years before 1914 (Butt, 1971).   Local-
authority transport and public-utility extensions,
for example, brought substantial betterment to
many landowners which they effectively exploited by
extolling the virtues of their sites in respect of
such amenities in local press advertisements. [11]
Where councils directly entered the land market to
restrain land prices in the face of mounting
population pressure, as in Sheffield, Liverpool,

51

Hereford, Birmingham, London and Glasgow,[12] an indirect effect was to undermine the profitability of working-class housing for the private sector.[13] The virtual cessation of this category of house-building, notably in the years immediately preceding World War I, led inevitably to more acute housing shortages. The imposition of building byelaws, with specifications regarding street width and building height, diminished the urban acreages available for building development in general and housing in particular. In the longer term, such council restrictions on building density towards the end of the century also reduced the rental capacity of sites and thus stimulated suburban land-extensive expansion rather than central land-intensive development.

In other respects the bout of Town Improvement schemes, though they cleared certain densely populated and insanitary areas, often stimulated the upward movement of prices for adjacent sites, as the displaced populace switched their rental bids to neighbouring housing zones and thereby introduced both higher land prices and, as a result, creeping deterioration and housing blight (Allan, 1965). In the worst cases, as in Leith and Glasgow, Improvement Trusts reduced absolutely the volume of available building land by themselves withholding it from the market, and also failed to comply with the requirements of the Cross and Torrens Acts by not providing alternative housing accommodation for residents displaced as a result of such 'improvement' operations.[14] As a result, landowners were endowed with further windfall gains as supply and demand for working-class housing moved further into disequilibrium.

Local-authority activities in another area also had important implications for the land market. Compulsory purchase and compensation procedures were cumbersome and costly, and ultimately acted as a deterrent to the extensive use of small area schemes to eradicate insanitary property. Private landowners, not surprisingly, withheld their property in the expectation of inflated prices. Existing sites thus became increasingly expensive as property owners, trustees and institutions recognized the advantage of black-mailing local authorities into exorbitant compensation payments and as a contemporary noted, 'the enormous expenses of the present method of determining compensation (the Land Clauses Act) ... is out of all proportion to the sum at stake' and owners used 'the threat of mulcting the Corporation in disproportionately heavy expenses to coerce them to give an excessive sum as compensation.'[15]   In ten Glasgow properties, for example, expenses averaged 23.9 per cent of the ultimate payment and in Greenock a levy of 4½d in the £ was necessary between 1877 and 1914 to pay for the improvement scheme and expenses associated with the arbitration award which had capitalized rental income at 14-25 years purchase 'for houses which ought not to have been inhabited at all.'[16]   In Glasgow compensation awards of from £4.10 to £11.10 per square yard for the Improvement Trust's acquisitions - equivalent to £21,700 to £55,600 per acre - inflated general land prices as higher expectations became written into landowners interpretations of the market.[17] Thus improvement schemes frequently undermined their own best intentions, for by inflating adjacent property prices and landowners' expectations of land prices, not to mention the burden distributed amongst all ratepayers, they prejudiced the availability and prices of sites and correspondingly influenced the insufficiency of privately produced working-class housing.

The system of land tenure may also have influenced the functioning of the land market, and the ultimate determination of the price of land. *A priori*, it may be suggested that freehold developments, lacking the reversionary gains of improvements at the expiry of the lease, would have led to a different land development strategy. Whitehand (1978, p.186) casts doubt on whether the type of tenure had much influence over the nature and intensity of land use.   He argues that: 'There is no reason to believe that *ceteris paribus* leasehold tenure gave rise to a lower intensity of use than freehold tenure', and Aspinall would also appear to concur with this conclusion.[18]

However, in 'another part of the island', namely Scotland, this was certainly not the case (Best, 1968).   Tenure and housing type were clearly connected[19], and this may have important implications for English housing studies.   The prevailing tenurial form in Scotland was the feuing system, certain aspects of which were akin to leasehold in perpetuity.   For the relinquishment of virtually all rights over the property, the landowner, in addition to the sum realised for the sale of the land, retained a position of 'superiority' which was a feudal legacy entitling him to exact a small sum (the 'feu duty') annually (Coull and Merry, 1971).   This had three particular implications for the land market.   First, with no

reversionary rights, the superior sought to
maximize the value of the site as he was liquidating
an irrecoverable asset, and would often tend to
withhold land from the market in the expectation of
higher land prices in the future. As one expert
described the position, '..the superior, after
feuing his land, is not interested in any increased
value of the land over and above the amount of the
feu duty, any increase in the value of the land
being participated in by the vassal alone'.[20] The
superior consequently endeavoured to increase the
selling price of his land to which the feu duty was
also related, and this in part explains the
existence of high-priced land on the outskirts of
Scottish towns. A second effect on the price of
land resulted from the fact that the level of the
feu was fixed absolutely. Thus to take account of
possible price inflation and land-value appreciation, the landowner was not encouraged 'to give his
ground for a reasonable feu-duty'[21] but to compen-
sate for reductions in the real value in the future
by creating particular onerous feu-duties in the
short term. The third influence was a product of
the process of 'subinfeudation'. The right to
exact an annual feu duty passed successively with
the transfer of land from owner to developer, and
then to the builder, residing ultimately with the
last buyer. As the right to extract such a payment was assured, the tendency was for vassals at
each stage of the chain to impose successively
higher feuing rates, so as to secure the original
amount for which they were personally liable, but
also to provide the basis of further income, and
thus the possibility of raising capital based on
this as security.

More obviously this tenure system also
affected the land market by imposing restrictions
on the uses to which a site could be put. The
'Feu Contract' in Scotland could prohibit certain
types of building or stipulate the quality and type
of materials to be employed and though similar
clauses could be applied in England, they were
relatively uncommon. By this means high density
working-class dwellings or commercial uses were, in
Scotland, restricted in an effort to control
estate development and enhance site values of as
yet undeveloped acreages.[22] Property transference
was thus a useful instrument for class confinement
and for residential and commercial segregation. In
constricting the operations of the land market it
was soundly criticized and as one authority stated,
'the one-sidedness of the present bargain between

54

superior and feuar is so obvious',[23] though
nothing of any consequence was achieved to rectify
the balance in the years before 1914.

In reality, therefore, the hypothetical,
perfectly functioning, land-use-allocating model as
postulated by Alonso (1964), Wingo (1961), Garrison
(1959) and others, was significantly amended.
Although the underlying rationale remains, in this
author's view, intact, empirical work demonstrates
that variants of the model are in fact the norm.
Institutional decisions, sub-optimal estate
development, imperfect information, the conflict of
long and short-term objectives, and the impact of
the tenure and tax systems seriously distorted the
allocation of land uses as suggested in the pure
version of the land-market model. Rent versus
accessibility is thus a useful first approximation,
but in the nineteenth century non-price distortion
effects were often stronger influences on the
eventual land-use patterns.

Empirical evidence: the land market in nineteenth-century England and Scotland

A number of qualifications have been advanced in
this brief survey of the operation of the land
market in the nineteenth century. It must now be
asked whether the land market has any historical
relevance and if so what form does it take?
Unfortunately, empirical studies are rather thin on
the ground. There are two main reasons for this:
first, there are shortcomings in the data sources,
such as the absence of a centralized property
register, and, second, the complex problem of
distinguishing the value attributable to buildings
and that portion ascribed to the land. Recently
Thompson's work[24] on the Yearbook of Auction Sales
at quinquennial intervals between 1892 and 1912 has
yielded further data on types of tenure, rental
values and the number of property transactions in
several London boroughs. This extension of his
earlier work (Thompson 1957) offers a useful
detailed complement to Vallis's (1972) more general
study. Richardson and Vipond (1970) have recently
paid some attention to changes in more modern sale
prices for Scottish properties. However the only
study to concentrate on land values in the long
run remains that of Singer (1941). Measurement by
Singer depended upon a definition of urban land
rent. This was 'the difference between income

derived from land when in actual or prospective use as an urban site and when used as agricultural land'. Income so obtained from an urban site was jointly composed of a land and a buildings element and the value of the land was determined by subtracting the building component from the total annual value. This 'residual' approach (Pribram, 1940), theoretically developed by the classical economists but never reduced to a practical calculation. (Kingsbury, 1946), can be seen as 'the amount remaining out of joint land and building income after income attributed to the building has been deducted ... any surplus on the land and building above that necessary to pay operating expenses, taxes, interest and depreciation on the building is properly assigned to the non-reproducible element in the combination - the land' (Hoyt, 1933, p.451).

In the unlikely event of obtaining a sufficient number of comparable sites whose ground rent element can be charted annually, the residual rent approach seems the only viable method of establishing an index of urban land rents on an annual basis. This is the approach adopted in the present study, which seeks to establish a minimum level of yearly increments to Scottish land values, though there seems no reason to doubt the general application of the trends and fluctuations to Britain as a whole. Unfortunately, as the data is only available at an aggregate Scottish or county level the average appreciation of urban land values must considerably underestimate the position in individual burghs.

Much the same procedures as those adopted by Singer are used, and the basic data source, the Schedule A tax assessment, is employed in obtaining a series for gross rentals (i.e. building and land values combined). However, three advantages over the Singer sources are apparent. First, Scottish reassessment intervals were yearly rather than triennially or less regularly, as in England and Wales (Stamp, 1922, Weber, 1960). Second, the actual net additional number of houses is also known[25] and hence Singer's restrictive assumption about the inelasticity of the building industry and the absence of substantial additions to the housing stock is avoided. Third, the customary system of yearly tenancies in Scotland, in contrast to the more flexible monthly lettings in England, meant that rent adjustments in the course of the year are not a complicating factor in the calculations.[26]

From the rateable value data provided by Inland Revenue officials gross rental data is

obtained.[27]   This, as Singer observes, gives 'a
figure for composite gross value' (Table 2.1, col.
(1)).   The difference between actual additions to
rental values (col. (1)) and the projections of
annual value based on the average rental of the
previous year and the number of existing new
houses (col. (2)) provides the increase in annual
value not due to building - in Singer's words, the
'unearned increment'.   This crude estimate of the
minimum addition to land values must, however, be
adjusted to take account of changes in the level of
building costs (col. (3)).   This is necessary
because if the land continued to be used for the
same purposes beyond the life of the existing
building, the higher building costs would eventually
be incurred and consequently the unearned increment
must be adjusted for this factor.   Only when this
has been completed are the building influences
removed from the gross rent with the 'residual'
rent reflecting annual increments to urban land
values (col. (4)).   As the data source distinguishes houses above and below £20 per annum
rateable value it is possible, following the same
procedures, to derive estimates of incremental land
values for, broadly speaking, working-class (col.
(5)) and middle-class housing (col. (6)).   Although
the £20 rateable value is rather high for some
labouring and artisan groups, it nonetheless may be
taken as a pointer to movements in the land component of different house rents.

The cyclical nature of land-value appreciation
and changes in urban ground rents has been stressed
by several writers (Hoyt, 1939; Duon, 1943;
Gottlieb, 1976). Limited British evidence from
Whitehead (1974) and Whitehand (1975), as well as
previously mentioned studies by Thompson and Vallis
provide an indication that in the late-nineteenth
and early-twentieth centuries British land values
also pursued a pattern of growth linked to the path
of housebuilding fluctuations.   Singer's own
evidence tends to bear this out, displaying a more
pronounced expansion in urban land rent in the
1890s for example, when the housebuilding sector
was noticeably buoyant.   Pribram (1940) also
stresses a theoretical under-pinning for the
tendency for urban land rents and building cycles to
be correlated, while he suggests that the rigidity
of land rents provided attractive hedging
opportunities for investments during periods of
falling prices.

The Scottish data in Table 2.1, cols. (4) and
(5), would lend further support to the argument for

Table 2.1(a): Scottish Urban Land Rents, 1864-1911
Total Housing Stock (£000s omitted)

					Urban Land Rent Increment	
Rateable Value of Houses (Gross Rent)	Earned Income (Income from Building)	Unearned Increment (adjusted for changes in building costs)	All Houses (3) as a % of (1)	R.V.<£20 p.a. (%)	R.V.>£20 p.a. (%)
(1)	(2)	(3)	(4)	(5)	(6)
1864					
65					0.45
66					0.37
67					0.67
68					0.19
69					1.20
70					0.72
71					0.44
72					0.22
73					0.10
74					-0.15
75					0.07
76	6026				0.66
77	6464	6402	62.3	0.96	-7.42	1.83
78	6844	6497	382.8	5.89	6.73	-0.70
79	7161	6566	685.9	9.57	9.87	0.77
80	7374	7265	110.9	1.50	1.71	0.77
81	7500	7473	26.1	0.34	0.85	-0.16
82	7658	7638	19.4	0.25	0.26	-0.01
83	7767	7758	8.5	0.11	0.00	0.75
84	7829	7854	-24.8	-0.32	-0.01	-0.06
85	7926	7935	- 9.1	-0.11	0.05	0.06

Table 2.1(a) (cont'd)

86	7968	8049	-81.4	-1.02	0.05	-1.75
87	8092	8063	28.0	0.34	0.77	-0.13
88	8255	8176	79.7	0.97	0.76	0.59
89	8365	8341	24.2	0.29	0.51	-0.26
90	8469	8455	14.1	0.17	0.57	-0.46
91	8598	8544	52.6	0.61	1.03	-0.37
92	8757	8608	87.3	1.00	1.10	-0.52
93	8912	8824	87.1	0.98	0.74	-0.50
94	9081	9033	77.7	0.86	0.98	-0.50
95	9279	9207	72.1	0.78	0.92	-0.53
96	9495	9301	184.2	1.94	1.88	-0.81
97	9766	9714	49.8	0.51	0.20	0.00
98	10079	9996	82.5	0.82	0.71	-0.87
99	10400	10269	123.9	1.19	1.33	-0.93
1900	10756	10624	134.3	1.25	1.55	-0.12
1	11074	10958	115.5	1.04	1.46	-0.10
2	11347	11261	86.1	0.76	1.10	-0.58
3	11662	11485	176.6	1.51	1.74	-0.24
4	12021	11836	185.0	1.54	2.06	-0.45
5	12310	12229	80.6	0.65	1.11	-0.45
6	12555	12449	105.7	0.84	1.21	-0.18
7	12781	12703	77.4	0.61	0.87	-0.22
8	12990	12904	85.9	0.66	0.89	-0.74
9	13128	13066	62.5	0.48	0.53	-0.04
10	13272	13184	88.1	0.66	0.53	0.25
11	13372	13308	63.3	0.47	0.84	-0.93

Source:
Columns (1) to (6) P.R.O. IR16, 1-135.

cyclical fluctuations in urban land rents, and
suggests that these were closely correlated with
variations in house building.   The aftermath of a
substantial upsurge in housebuilding in the mid-
1870s appears to have considerably inflated land
values, though with a lag of up to three years.
There also appears to be an interval before urban
ground rents adjusted to the depressed building
conditions of the 1880s, although in that decade
the generally unfavourable house building market was
matched by a stagnating level of urban ground rents,
notably in the years 1883-6.   Indeed, static land
values as reflected in the virtual absence of
additions to ground rents in the working-class
section of the land market, were accompanied by
small percentage reductions overall in these years.
The resumption of building, albeit temporarily, in
1887-9 and then more permanently in the early years
of the 1890s gave further encouragement to the land
market as reflected in the increments to land rents
in those years.   Housebuilding, if not exactly
flowering after the 1903 boom, did hold up well
until 1906 in most burghs and somewhat later in
towns such as Stirling and Dunfermline, while urban
land rents tended to mirror this pattern,
particularly in the working-class section of the
market.
        The value of existing land usage (which is to
be distinguished from the potential development
value) in Scotland, rose on average by 1.09 per
cent per annum between 1876 and 1911, though if
the peculiar years of the late 1870s are excluded,
the average annual land-value increase was 1.68
per cent (Table 2.1).   For sites occupied predominantly by working-class housing the average
land-value increase was slightly higher, 1.10 per
cent per annum over the period 1876 to 1911 and,
more typically, averaged 0.91 per cent per annum in
the years 1880 to 1911.   It should be remembered,
however, that this average figure is for both
urban and rural housing sites and the likelihood is
that it underestimates the urban element, possibly
by a considerable margin in some burghs.   This
would appear to be the only explanation of the
annual land-value decrease of 0.22 per cent for
sites occupied by properties rated at more than £20
per annum.   In the context of continued urban
expansion it seems unreasonable to suggest that
better types of housing should be situated on sites
which were declining in value, but it does appear
that decreasing land values in the rural housing
sector rated above £20 per annum - as in the areas

Table 2.1(b): Scottish Urban Land Rents, 1864-1911

Scottish Indices (1900 = 100)

	Housebuilding (1900-09 av=100)	Building Costs	Rents
	(7)	(8)	(9)
1864		59.9	
65		61.7	
66		67.2	
67		67.0	
68		66.8	
69		67.7	
70		67.0	
71		67.6	
72		71.7	
73	168.2	77.7	
74	194.6	82.2	
75	211.5	88.3	79.8
76	247.2	90.7	81.4
77	193.3	89.5	84.5
78	78.5	81.0	86.1
79	52.3	70.3	88.5
80	111.0	69.3	89.0
81	71.7	71.4	89.2
82	74.5	72.1	89.5
83	57.7	75.0	89.3
84	78.9	75.1	89.0
85	76.7	75.1	89.1
86	94.4	75.1	88.6
87	105.5	76.4	89.2
88	97.0	76.4	89.5
89	96.0	79.3	89.6
90	72.4	81.6	89.8
91	76.8	83.8	90.7
92	87.6	85.3	91.6
93	107.3	86.2	92.5
94	124.0	87.1	93.3
95	125.7	87.2	94.0
96	153.4	91.7	94.4
97	156.5	95.5	96.1
98	168.1	95.7	97.8
99	138.4	101.5	98.6
1900	93.6	100.0	100.0
1	121.4	100.4	100.7
2	132.2	100.2	101.5
3	125.7	100.2	103.6
4	116.3	100.2	104.9
5	113.0	100.2	105.1

61

Table 2.1(b) (cont'd)

	Housebuilding (1900–09 av=100)	Building Costs	Rents
	(7)	(8)	(9)
1906	106.0	100.2	106.4
7	69.6	100.2	106.6
8	59.9	100.4	107.7
9	62.9	100.2	107.7
10	51.9	100.2	108.7
11	35.1	100.4	109.0

Source:

Columns 7 to 9 R.G. Rodger (1975)'Scottish Urban Housebuilding 1870-1914', Unpublished Ph.D. Thesis, University of Edinburgh.

bordering the Solway coast where rural depopulation was reflected in an absolute decline in the number of occupied houses - were proceeding sufficiently rapidly to neutralize land-value increases in peripheral areas of the major towns.

Scottish evidence thus seems to verify Mill's (1875, p.284) conclusion that, 'the land of the world ... constantly increases in value'. The operation of the land market required most Victorians to pay a steadily increasing weekly amount to cover the ground rent and, as demand and supply influences were only rarely in equilibrium, the price of land pursued an upward direction at an uneven pace.

The absolute level of land prices was crucial to the development of working-class housing. As a prominent Glasgow builder explained: "Ground is the builder's first requisite. It is fixed in regards position and limited in area, and in the neighbourhood of large towns becomes dear ... its price has a considerable influence in determining the rents payable by the occupants."[28] The relationship of rents to ground rents was discussed by another official enquiry[29] and effectively forged the link between market prices for land and housing. The report noted that'...although the portion of the rent of a tenement house which is attributable to the value of the land may be small, this does not imply that the cost of land is a negligible factor in the provision of housing accommodation; because the houses have been crowded together for the very purpose of reducing the element of land value as

62

much as possible.'   This was a reiteration of the
contemporary view that the land market was 'a
primary factor making for the congested development
of towns' with the result that 'the ring of high
price land surrounding the town hems in the population like a wall.'[30]
   The combination of urban expansion and the
operation of the Victorian labour market (Jones,
1971; Treble, 1978) exerted considerable pressure
through the price of land on the quantity and
quality of accommodation which labourers and
artisans could afford.   Consistent with the theoretical formulation of the land market model a Royal
Commission report noted that the manual worker
'pays an infinitely greater sum for his land on
which his house stands than the man in the well-to-
do residence.'[31]   Indeed for proximity to his workplace the breadwinner required his dependants to
pay dearly.   High-density working-class housing on
the margins of the industrial and commercial areas
imposed environmental conditions which were
seriously injurious to health and welfare as was
eventually recognized in various statutes and bye-
laws, and it was this premium on accessibility to
the workbench which imposed considerable strains on
the other members of his household.   The deleter-
ious effect on family health was explicit in the
view of one report:'...there is the loss to consider
of fresh air, of sunlight, of elbow room and
privacy - factors in life which however much or
little they may affect the wage earner are of vital
importance to the mothers and children of the
nation.'[32]
   Furthermore, low-amenity, high-density housing
influenced the physical deficiencies of children
(Mackenzie, 1913), a topic which, as the nineteenth
century drew to a close, the nation felt it could
no longer completely ignore.   Various studies
examined the relationships between health and
housing, and pointed to high land values in city
centre districts, the preserve of the working
classes, as one of the fundamental causes of poor
health.   One such study showed that fourteen-year-
old boys from 'model' houses in Port Sunlight were
at least 4 - 7 inches taller and 13-27 lbs. heavier
than their equivalent age group in Liverpool Council
Schools, and were approximately equal in physique
to their contemporaries who attended Higher Grade
schools in 'well-to-do' areas of Liverpool.   In
Glasgow, too, the physical development of 72,857
children aged 5-18 was closely correlated with their
housing conditions and domestic background.   The

study concluded "it cannot be an accident that boys from two-roomed houses should be 11.7 lbs lighter than boys from four-roomed houses and 4.7 inches smaller. Neither is it an accident that girls from one-roomed houses are on average 14 lbs lighter and 5.3 inches shorter than girls from four-roomed houses."[33]

The overcrowding problem in Scotland was in fact particularly severe. One-roomed houses accounted for 12.8 per cent of the housing stock compared to 3.2 per cent in England and Wales; two-roomed houses in Scottish towns of over 2000 population accounted for 56.8 per cent of houses as compared to 8.3 per cent in England and Wales. The position in many Scottish burghs was of course much worse and in Kilmarnock, Dundee and Coatbridge for example respectively 67.7 per cent, 69.9 per cent and 78.7 per cent of the residents lived in houses having no more than two rooms. Consequently almost half (49.6 per cent) of Scots were housed in one or two-roomed properties; south of the border it was 7.1 per cent.[34]

These housing conditions - 'one rather striking feature of urban Scotland'[35] - were no haphazard consequence of consumer preference, 'clannishness' or climatic conditions. Nor is a narrow nationalist debating point at issue. Urban Scotland simply offered an example, *in extremis,* of the result of very high-priced central land, admittedly in conjunction with lower levels of real wages. These, in 1905, were on a par with the agricultural counties of Southern and Eastern England, but were between 11.5 per cent and 19.5 per cent below levels operating in the Midlands, Lancashire, Yorkshire and London.[36] Thus unskilled labourers in Middlesbrough, Ipswich or Wolverhampton could, in 1902, commonly obtain three roomed houses for rents of between 3s. 6d. and 4s. 6d. per week. In Glasgow or Clydebank, the same category of worker would normally have been able to secure two rooms only at that rental level, and it was hardly surprising that overcrowding was correspondingly higher (W. Thompson, 1903).

The Board of Trade estimated that Scottish rents were some 10.1 per cent to 27.5 per cent above other areas in the United Kingdom, including London, and one contributory factor in this was the price of land. With typical understatement the conclusion of Whitehall was that '...the rent index numbers for the Scottish towns were somewhat high as compared with those for English and Welsh towns ...'.[37] The average floor-space rent in pence (d)

per square foot according to whether the housing was
of cottage, flatted cottage or tenement style was in
England, respectively, 8.49d, 6.92d and 8.25d. In
Scottish burghs, these figures were higher in each
case, namely 8.62d and 10.20d per square foot.[38]
More significantly the dominant housing type,
cottage or flatted cottage in England and tenement
in Scotland, showed a marked divergence in rental
levels - a divergence which, if the Board of Trade
estimates were accurate, was narrowing slightly with
time and which therefore could have only been wider
in the late-nineteenth century. Table 2.2 shows
average floor space rentals in various English and
Scottish towns, according to the dominant housing
form. Although such comparison between towns is

Table 2.2: Average Floor Space Rent: Various English and
Scottish Towns, 1911

English boroughs: Rent per sq.ft. (d)		Scottish burghs: Rent per sq.ft. (d)	
Liverpool	10.24	Glasgow	11.29
Sunderland	10.03	Paisley	11.22
Barrow	8.25	Govan	11.00
Stockton	7.34	Partick	10.49
Portsmouth	6.02	Dundee	10.18
South Shields	5.87	Edinburgh	9.99
Chatham	5.54	Leith	9.26
		Aberdeen	8.20

Source: Royal Commission on the Housing of the Industrial
Population of Scotland, Rural and Urban, 1917.
Evidence and Appendices (H.M.S.O. 1921), Appendix
CLXVII.

problematical, and only limited evidence of this
type is available, the discrepancies in floor-space
rentals payable in, for example, Liverpool and
Glasgow, or South Shields and Govan are significant.
Comparison of rental payments according to the
amount of accommodation purchased are further complicated by the variable level of building costs for
labour and materials between cities. However the
conclusion to be drawn from Table 2.3, which
presents the land-costs component in relation to
total rental payments for buildings of a similar
design and construction in London and Glasgow, is
that the land element was appreciably higher in
Glasgow.
London County Council paid an average of
£3,580 per acre for the thirty-eight acres of land

Table 2.3: Ground Rents as a Percentage of Total Rents
(Comparable Block Dwellings in London and Glasgow)

Glasgow Improvement Trust

Saltmarket (1890)	15.73%
Norrin Square (1895)	12.01%
Kirk Street (1896)	14.26%
St. James' Road (1897)	12.21%

Glasgow Dwellings Company

Cathedral Court (1892)	9.50%

London East End Dwellings Co.

Lolesworth Buildings (1887)	7.97%
Katherine Buildings (1885)	6.75%

London Common Trust

Brandon Street (1897)	7.20%
Page's Walk (1895)	7.49%

Source: J. Mann (1898),'Better Houses for the Poor - Will
they pay?' *Proceedings of the Royal Philosophical
Society of Glasgow,* 30, 83-124

purchased between 1893 and 1901 for the purpose of
municipal housebuilding (W. Thompson, 1903).   This
average figure was hardly likely to understate the
price of land in the metropolis and as the council
had to compete in the open market for land it
probably is a fairly accurate reflection of London
land prices.   English provincial councils paid an
average figure of £2,730 per acre for land which
they used for city centre improvements in the same
years.   By contrast, the prevailing land market
prices in Glasgow and Edinburgh were approximately
£4,000 per acre over the period in question.[39]
Thus in the 1890s Glasgow building land cost upwards
of 30s. per square yard; in a comparable English
town such as Liverpool, a price of 22s.6d. to 24s.
per square yard was customary.
     The higher cost of land in Scotland, produced
principally by the peculiar feudal tenure system,
combined with real wages substantially below comparable English centres to produce a lower threshold
to effective demand for housing space.   Urban
living conditions were consequently harsh by
comparison.   High land prices thus produced a distinctive residential pattern of high-density housing

associated with which were acute problems of overcrowding and environmental deprivation (W. Thompson,
1907; Russell, 1877, 1887).

The demarcation of residential areas was, of
course, also apparent in English boroughs. Just as
there were clearly defined middle-class suburbs such
as Morningside and Murrayfield in Edinburgh and the
western extensions in Glasgow, so there were Alderley
Edge and Wilmslow in Manchester, Edgbaston in
Birmingham, Clifton, Lotham and Redland in Bristol
and similar concentrations in Sheffield, Nottingham,
London and indeed most urban areas (Burnett, 1978;
Simpson and Lloyd, 1977).   Middle-class disaffec-
tion with the city centre and its associated
environmental hazards combined, in the face of
mounting working-class pressure on land prices, to
push such locations beyond their means. The pricing
mechanism acted in the land market as a filter in
the process of segregation, each gradation of price
operating as a settling pool for a reasonably well-
defined social group.

In addition, middle-class villa building tended
to be more evenly distributed over the course of
the building cycle.[40]   With relatively cheaper land
in the downswing phase of the building cycle, such
as the 1880s, a centrifugal influence was contributed
to the housing pattern and zones of middle-class
housing developed when effective mass demand was
powerless to prevent it.   The economic weakness of
working-class groups, and their inability during
trade recessions to occupy and expand into certain
parts of the city, offered higher income groups the
opportunity to do so uncontested as desirable properties could be secured at relatively low prices
due to the cyclical stagnation in land and building
costs.   One result of this process was to solidify
a band of superior residential properties which
acted as a barrier, segregating middle- and working-
class inhabitants, and which effectively pre-empted
subsequent working-class intrusions into the clearly
demarcated villa zone.[41]

Visually, too, the Victorian city owed something to the functioning of the land market.
Working-class properties tended to be constructed in
a few concentrated years at the peak of the building
cycle.   Consequently the need for economy and the
historically limited period of building combined to
produce homogeneity of style (Tarn, 1973). An
architectural uniformity was therefore stamped on
terraced and tenement housing, which was partially
the product of cyclically-related increases in land
prices.   In contrast, higher and more permanent

incomes, lower land prices, differing intensities of land and other building costs and a more even distribution of housebuilding throughout the course of the cycle, allowed middle-class residents to indulge their diverse tastes in the design and construction of their nineteenth-century suburban homes. Institutional building was also cyclically concentrated because of its land extensive requirements, and this contributed further architectural stereotypes to the urban townscape. A counter-cyclical bout of Scottish institutional building between 1905 and 1914 in which many towns undertook the construction of baths and wash-houses, libraries, museums, municipal workshops and a host of other civic building, including the erection of local authority office accommodation, offers just one example of the architectural uniformity which tended to emerge in the period preceding the First World War (Rodger, 1979).

Conclusion

The proliferation of public health statutes, the development of limited housing improvement schemes and their ultimate translation into town planning projects, the legislative intrusion of central government, supplemented by the adoption of local byelaws, all successively influenced and distorted the operation of the land market. From a relatively unfettered position in the mid-nineteenth century, with the price mechanism wholly responsible for clearing the market and allocating land to its various uses, the market became increasingly adulterated by institutional interference, albeit undertaken for enlightened, though sometimes misguided, reasons. As such, it was often precisely those agencies acting in the name of reform which contributed most to the progressive breakdown in the Victorian land market in its purest form, for in taking decisions about parks, drainage, street width, building height, health risks and a host of other proliferating interests the land market was both directly and indirectly affected. Windfall gains accrued to private landowners and transport and public utility developments further supplemented the process by which public decisions influenced private individuals, notably the interests of landowners. Thus the intervention of the state in the land market both explicitly and implicitly changed existing and potential land-use values. Anomalies and inconsistencies were introduced to the ordered and unrestricted land market, and the estate development strategies of landowners were accordingly determined more by the decisions of the council than by the price mechanism in its unrestricted form (Parker, 1965).

Not surprisingly, these changes prompted calls for land-value taxation, for reform and betterment levies, notably from Henry George (1881). This tacit acknowledgement of the problem of land values, as shown by the multiplicity of tracts dealing with the subject between 1890 and 1910[42], is indicative of the frustrated workings of the pricing mechanism in the land market and the resort to, and inconsistencies created by, land-use allocation by other means. That there were extensive deliberations over the betterment taxation issue in the two decades before the war is further evidence of the consternation felt over the seemingly intractable problem of working-class housing, ascribed mainly to a malfunctioning of the land market. Curiously, the land market was expected to achieve an equitable distribution of housing needs. It was, however, merely another factor market in which welfare considerations did not play a role. Thus, just as society sought to impose controls in the labour market, it also attempted to redress what it perceived as inequitable land allocations for housing the poorest groups. In fact, it was not so much that the land market had malfunctioned, more that the market-clearing solution arrived at by the price mechanism produced an increasingly unacceptable outcome. In the vain search for some panacea to ease the mounting housing and environmental problems within nineteenth-century cities the operations of the market were increasingly circumscribed. Consequently, the mechanism of the market was upset and the working-class housing problem further exacerbated.

Notes

1. *R(oyal) C(ommission) on the housing of the industrial population of Scotland, rural and urban*. Cd. 8731, 1917, Report, para. 1545, 1573; Minority Report, para. 104-5; Evidence (H.M.S.O. 1921) Mickel, Q. 21893; *G(lasgow) M(unicipal) C(ommission) on the housing of the poor* (Glasgow, 1904), Eadie, p.336; Menzies, Q. 3321

2. H.J. Dyos, 'Agenda for Urban Historian' in Dyos (ed.) (1968), *The Study of Urban History*, Arnold, London, gives an extensive bibliographical list of relevant studies. For example his footnotes 59, 106, 125.

3. *GMC*, McGillivray and Fraser, p.593

4. *R.C. 1917*, Nettleford, Q. G2755; Ross, Q. 33719

5. Ibid. and *GMC*, McGillivray and Fraser, p.593

6. Ochojna, A.D. (1974) 'Lines of class distinction: an economic and social history of the British tramcar, with special reference to Edinburgh and Glasgow,' Unpublished Ph.D. thesis, University of Edinburgh, ch.5

7. *Presbytery of Glasgow report of commission on the housing of the poor in relation to their social condition,* (Glasgow, 1891), p.26 "There seems no doubt that landlords and speculators are encouraged by freedom from taxation to hold building lands in the neighbourhood of great cities for a high price, and there can be little doubt that this policy tends to artificially enhance the value of building areas, and tends to overbuilding and overcrowding within cities, as well as the enhancement of rent."

8. *S(cottish) L(and) E(nquiry) C(ommittee),* (London, 1914), Report, p.484

9. For example, English league for the taxation of land values, land values Parliamentary group, committee for the taxation of land values, Yorkshire land values committee.

10. *R.C. 1917*, Report, para. 1706

11. *The Scotsman,* 21st July, 1883. This was just one of many adverts which carried the phrase 'bus passes door' or something similar.

12. *R.C. 1917*, K. Taylor, Q. 24337, 24378; Turton, Q. 24741-3; Brodie, Q. 41936; Wilson, Q. 44039; W. Smith, Q. 41557, Aldridge, App. CLX1X

13. *GMC*, Binnie, Q.6551, 6554, 6878-94, 6920, 7029, 7171; Nisbet, Q.1490, 1497

14. *R(oyal) C(ommission) on the housing of the working classes,* Cd. 4409, 1885, Laing, Q.20431-532; Russell, Q.19324-36, Morrison, Q.19321-3

15. *R.C. 1917*, Report, para. 1652-3. Brackets added.

16. *R.C. 1917*, Report, para. 1667-8; A. Evidence of Lindsay. Q. 23299, and App. LXXXII; *R.C. 1855,* Report, p.6

17. *GMC*, Eadie, p.337, charges the local authority with considerable responsibility for increasing land prices: "The extraordinary prices asked by the Trust for their land upset all preconceived notions as to the value of ground in the city. Property values were revolutionised, wild speculation set in, values and rents were jerked up all round..."

18. Aspinall, P.J. 'The evolution of urban tenure systems in nineteenth-century cities' (unpublished typescript)

19. *SLEC*, p.293, "It is very probable also that the preponderance of three and four storeys in height in Scotland as compared with England is attributable in part at least to these high ground rents."

20. *R.C. 1917*, para. 1546

21. *SLEC*, p.309-10

22. Simpson, M.A. (1970) 'Middle class housing and the

growth of the suburban communities in the East end of
Glasgow, 1830-1914', Unpublished B.Litt. thesis, University of
Glasgow, pp.22-34

23. *SLEC*, Report, p.lii

24. Thompson, F.M., (1977), 'Computer Aided Analysis of
the London Property Market, 1892-1912, *S.S.R.C. Report*

25. P(ublic) R(ecord) O(ffice), IR 16, 1-135

26. *Departmental committee on house-letting in Scotland*,
Cd. 3715, 1907, Report, p. 3-6, GMC, Donald, Q. 10484; Mann,
Q.9379

27. PRO, IR 16, 1-135

28. *GMC*, Eadie, p.336

29. *SLEC*, p.326

30. *Ibid.*, p.390, 541

31. *R.C. 1917*, Campbell, Q. 18774

32. *R.C. 1917*, Minority Report, Para. 116

33. *Report on the physical condition of children in
Glasgow*, H.M.S.O. 1907; Lord Leverhulme, (1918) *Six Hour Day
and Other Industrial Questions*, Allen and Unwin, London

34. *Census of Scotland*, 1911; *Census of England and
Wales*, 1911

35. *SLEC*, p.283

36. *Report of an enquiry by the Board of Trade into
working class rents, housing and retail prices etc.* Cd. 3864,
1908, p. xi

37. Ibid., p. xxxviii

38. *R.C. 1917*, Appendix, CLXVII

39. *R.C. 1917*, para. 592

40. Rodger, R.G. (1975) 'Scottish Urban Housebuilding,
1870-1914'. Unpublished Ph.D. thesis, University of
Edinburgh

41. Aberdeen town planning register, 1880-1914, Post
Office directories, 1848/9-1914: Glasgow Dean of Guild court
registers, 1873-1914, Pollokshaws register of plans to 1912,
East Pollokshields Dean of Guild court register of plans and
Pollokshields housing petitions, 1880-91, Glasgow Post Office
directories, 1828/9-1914; Greenock Post Office directories
1853/4-1914; Edinburgh Dean of Guild court registers, 1880-
1914 and Post Office directories 1879-1914

42. Ward, S.B. (1976) 'Land reform in England, 1880-
1914'. Unpublished Ph.D. Thesis, University of Reading. The
bibliography offers ample testimony on this point.

References

Allan, C.M. (1965) 'The genesis of British urban re-development with regard to Glasgow, *Economic History Review*, 18, 598-613

Alonso, W. (1960) 'A theory of the urban land market', *Papers and Proceedings of the Regional Science Association*, 6, 149-57

Alonso, W. (1964) *Location and land use*, Harvard University Press, Cambridge, Mass.

Beckmann, M.J. (1969) 'On the distribution of urban rent and residential density', *Journal of Economic Theory*, 1, 60-7

Best, F.A.A. (1968) 'The Scottish Victorian City', *Victorian Studies*, 11, 329-58

Burgess, E.W. (1925) 'The growth of the city: an introduction to a research project' in R.E. Park and E.W. Burgess, (eds.), *The City*, Chicago University Press, Chicago, pp.47-62

Burnett, J. (1978) *A social history of housing 1815-1970*, David and Charles, Newton Abbot

Butt, J. (1971) 'Working class housing in Glasgow 1851-1914' in S.D. Chapman (ed.) *The History of Working Class Housing*, David and Charles, Newton Abbot, pp.57-92

Cannadine, D.N. (1977) 'Victorian cities: how different?', *Social History*, 4, 457-82

Clarke, C. (1965) 'Land taxation: lessons from inter-national experience', in P. Hall (ed.), *Land Values*, Sweet and Maxwell, London pp. 126-44

Coull, J.W. and Merry, E,W. (1971) *Principles and Practice of Scots law*, Butterworths, London

Davie, M.R. (1961) 'The Pattern of Urban Growth', in G. Theodorson (ed.) *Studies in human ecology*, Harper Row, New York, pp.77-92

Dickinson, H.D. (1969) 'Von Thünen's economics', *Economic Journal*, 79, 894-902

Duon, G. (1943) 'Evolution de la Valeur Venale des Immeubles Parisiens', *Journal de la Societé de Statistique*, 84, 169-92

Ely, R.T. and Wehrwein, G.S. (1940) *Land Economics*, Macmillan, New York

Garrison, W.L. *et al.* (1959) *Studies of highway development and geographic change*, Washington U.P., Seattle

George, H. (1881) *Progress and Poverty*, Routledge and Kegan Paul, London

Gottlieb, M. (1976) *Long swings in urban development*, N.B.E.R., New York, pp.149-55

Hallett, G. (1977) *Housing and land policies in West Germany and Britain*, Macmillan, London

Harris, C.D. and Ullman, E.L. (1945) *The nature of cities*, Annals of the American Academy of Political Science

Hay, J.R. (1975) *The origins of the Liberal welfare*

reforms 1906-14, Macmillan, London

    Hoyt, H. (1933) *One hundred years of land values in Chicago,* Chicago University Press, Chicago

    Hoyt, H. (1939) *The structure and growth of residential neighbourhoods in American cities,* U.S. Government Printing Office, Washington, (D.C.)

    Hurd, R.M. (1903) *Principles of city land values,* The Record and Guide, New York

    Jones, G.S. (1971) *Outcast London,* Clarendon Press, Oxford

    Kain, J.F. (1962) 'The journey to work as a determinant of residential location', *Papers and Proceedings of the Regional Science Association,* 9, 137-60

    Kellett, J.R. (1969) *The impact of railways on Victorian cities,* Routledge and Kegan Paul, London

    Kingsbury, L.M. (1946) *The economics of housing,* Kings Crown, New York

    Lowry, I.S. (1967) *Seven models of urban development,* Rand Corporation, Santa Monica, California

    MacKenzie, W.L. (1913) 'The child of the one-roomed home', *Journal of the Child Study Society,* 6, 101-4

    Mann, J. (1898-9) 'Better houses for the poor - will they pay?', *Proceedings of the Royal Philosophical society of Glasgow,* 30, 83-124

    Matthews, R.C.O. (1959) *The trade cycle,* Cambridge University Press, Cambridge

    Mill, J.S. (1875) *Dissertations and discussions,* Vol. IV, Longmans, London

    Parker, H.R. (1965) 'The history of compensation and betterment since 1900' in P. Hall (ed.) *Land Values,* Sweet and Maxwell, London, pp. 53-70

    Pribram, K. (1940) 'Residual, differential and absolute ground rents and their cyclical fluctuations', *Econometrica,* 8, 62-78

    Ratcliffe, R.U. (1949) *Urban Land Economics,* Greenwood Press, Westport, Connecticut

    Richardson, H.W. and Vipond, J. (1970) 'Housing in the 1970s', *Lloyds Bank Review,* April, pp. 1-14

    Rodger, R.G. (1979) 'Building cycles and the urban fringe in Victorian cities', *Journal of Historical Geography,* 5, 72-8

    Russell, J.B. (1877) 'Local vices of building as affecting the death rate', *Sanitary Journal for Scotland,* pp. 407-8

    Russell, J.B. (1887) 'The House in Relation to Public Health'. *Transactions of the Insurance and Actuarial Society of Glasgow,* 5, 123

    Saul, S.B. (1962) 'Housebuilding in England, 1890-1914', *Economic History Review,* 15, 119-37

    Segal, D. (1977) *Urban Economics,* Richard D. Irwin Inc., Illinois

    Simpson, M.A. and Lloyd, T.H. (1977) *Middle class housing*

in Britain, David and Charles, Newton Abbot

Singer, H.W. (1941) 'An index of urban land rents and house rents in England and Wales, 1845-1913', *Econometrica,* 9, 221-30

Sjoberg, G. (1976) 'The nature of the pre-industrial city, in P. Clarke (ed.), *The Early Modern Town,*Longmans, London, pp.43-53

Stamp, J.C. (1922) *British incomes and property,* King, London

Stigler, G. (1966) *The theory of price,* Macmillan, New York

Sutcliffe, A. (1970) *The autumn of central Paris. The defeat of town planning,* Arnold, London

Tarn, J.N. (1973) *Five per cent philanthropy,* Cambridge University Press, Cambridge

Taylor, T.G. (1949) *Urban Geography,* Methuen, London

Thompson, F.M.L. (1957) 'The land market in the nine-teenth century', *Oxford Economic Papers,* 9, 285-308

Thompson, W. (1903) *The housing handbook,* National Housing Reform Council, London

Thompson, W. (1907) *Housing up-to-date,* National Housing Reform Council, London

Treble, J.H. (1978) 'The seasonal demand for adult labour in Glasgow, 1890-1914', *Social History,* 3, 43-60

Vallis, E.A. (1972) 'Urban land rent and building prices, 1862-1969', *Estates Gazette,* 222, 1015-9

Weber, B. (1960) 'A new index of house rents for Great Britain, 1874-1913', *Scottish Journal of Political Economy,* 7, 233

Wheaton, W.C. (1974) 'A comparative static analysis of urban spatial structure', *Journal of Economic Theory,* 9, 223-37

Whitehand, J.W.R. (1975) 'Building activity and the intensity of development at the urban fringe: the case of a London suburb in the nineteenth century', *Journal of Historical Geography,* 1, 211-24

Whitehand, J.W.R. (1977) 'The basis for an historico-geographical theory of urban form', *Transactions of the Institute of British Geographers,*N.S.2, 400-16

Whitehand, J.W.R. (1978) 'The building cycle and the urban fringe in Victorian cities - a reply', *Journal of Historical Geography,* 4, 175-91

Whitehead, C.M.E. (1974) *The U.K. housing market: an econometric model,* Saxon House, Farnborough

Williamson, A.M. (1917) *The influence of housing on health,* Privately published

Wingo, L. (1961) *Transportation and urban land use,* Harvard University Press, Cambridge, Mass

Wohl, A.S. (1977) *The eternal slum,* Arnold, London

Youngson, A.J. (1966) *The making of classical Edinburgh,* Edinburgh University Press, Edinburgh

Chapter 3

THE INTERNAL STRUCTURE OF THE HOUSEBUILDING INDUSTRY IN NINETEENTH-CENTURY CITIES

Peter J.Aspinall

Introduction

Eighteen years ago Saul (1962-63) suggested that the
observed wide differences in the degree of speculative building from one area to another in the late-
nineteenth century might have arisen in part from
variations in the size structure of building firms
and the financial and other resources which they
employed.   For the first time historians were
explicitly encouraged to turn their attention from
the construction of complex economic models of the
Atlantic economy to investigations of the internal
structure of the building industry, in order to
achieve a greater understanding of how the housing
system functioned.   Yet most research on nineteenth-
century building has continued to add to our knowledge of the aggregate structure of the industry
nationally.   This state of affairs derives largely
from the divergent approaches of historians and
economic theorists.   While the former have shown
a marked reluctance to draw upon economic theory,
econometricians have similarly failed to utilize
the ever-growing body of quantitative and statistical evidence about the past (Lewis, 1965).   Work in
both fields has suffered.   The explanatory content
of a good many econometric investigations is
weakened by the unsatisfactory nature of the
behavioural assumptions of equations employed in the
construction of theory and by formulations at the
wrong level of aggregation.   Historians, for their
part, have tended to account for variance in terms
of the unique and particular in their pursuit of
explanations for historical events.
       In consequence, our knowledge of house-
building in the nineteenth century has suffered
from a high degree of compartmentalization.   Perhaps
most is known about secular and spatial fluctuations

in the output of the industry, much less about the
size and cost structure of the factor inputs, and
surprisingly little about the structure of the firms
themselves.   The now considerable volume of work on
suburban estate development has provided us with a
wealth of information on the events and decisions
which influenced the conversion of rural land to
urban use, but most of these studies have assigned
little separate consideration to the vast number of
builders who played a major part in bringing about
this transition.   While insights are occasionally
gained into how individual firms reached their
decisions to build, nothing is said of the manner in
which, collectively, the decisions of a whole
'population' of firms responded to changes in market
conditions in terms of their size and organizational
structure.   Only in the last few years has the role
of builders and developers, and their responses to
market trends, come under systematic investigation.
Yet the opportunities for research in this field are
considerable, and the following paragraphs suggest
some potentially fruitful lines of enquiry.

The scope for future research

Firstly, the degree of firm-size differentiation in
the nineteenth-century urban system is still open to
speculation.   The pioneering work of Dyos (1961,
1968, p.652) on the analysis of the District
Surveyors' monthly returns for south London and the
Borough of Camberwell, provided us with our first
detailed picture of the size-array of building firms
in a large part of a major city showing that '...the
large firms of "master builders" or general contractors were outnumbered by the small by at least
ten to one...'.   Studies of other areas of London
confirm Dyos's findings (Imray, 1966; Sheppard, 1971;
Crossick, 1978) yet the burgeoning Victorian cities
and the major manufacturing regions in the provinces
have been almost entirely neglected.   One wonders if
the size structure of the house-building industry
had a standard form everywhere or whether the
hierarchical structure of the urban system, of itself, produced a marked individual variance from
city to city.   Dyos (1968) speculated that the
logistics and finances of house-building did differ
from place to place,but added that too little
detailed research had been done on the operations of
builders up and down the country to confirm his
impressions, while Sutcliffe (1972, p.44) wrote that
'...it remains unclear whether provincial towns were

dominated by very large firms to the same extent as
London was by the end of the century when, according
to Dyos, seventeen firms were building over 40% of
London's houses'.   Analysis of details of employers
in the 1851 census suggests that they were not.
Only in the north-western counties, one of the most
highly urbanized areas, did the relative importance
of firms employing 50 or more men approach metropolitan levels with 8 or 9 per cent of all employers
falling into this category.   In the other regions
firms of this size accounted for only 3 per cent
to 5 per cent, while in every region those employing
under twenty persons accounted for at least 65 per
cent of the total and in the eastern counties over
90 per cent (Figure 3.1).   The variations may have
arisen from differences in the scope for speculative
building offered by the land market, but such
explanations must await further research on the size
structure of the provincial industry.   It is also
possible that the census returns themselves contain
regional inconsistencies in both accuracy and
terminology.

Secondly, most recent studies of the size
structure of the building industry are static in
method and tell us nothing about long-run trends of
change in structure (Daunton, 1977).   Nor do contemporary studies of size distributions provide useful guidance, an enquiry into the dynamics of small
firms finding '...virtually no information on the
changes taking place within small firms sectors in
terms of the numbers of new entrants, the age of
established firms, reasons for liquidation and so
on.' (Committee of inquiry on small firms, 1971,
p.6).   The need for this research stems from the
overriding necessity to relate the findings of
empirical studies of firm size to the growing understanding of the shape and changes of manufacturing
size distributions.   The increasing attention which
this subject has received from economists interested
in market concentration has revealed that such
configurations are almost always highly skewed in a
positive direction.   Attempts to explain these
findings have invariably assumed that the basic
causal mechanism was the shape of the long-run
average cost curve.   In homogeneous industries
this has generally been regarded as U-shaped,
because the readiness with which demand can shift
between the products of firms yields a distribution
in which most of such firms would approximate the
optimum size.   In heterogeneous industries, however,
Wedervang (1964) has shown that the influence of the
long-run average cost curve is modified by a

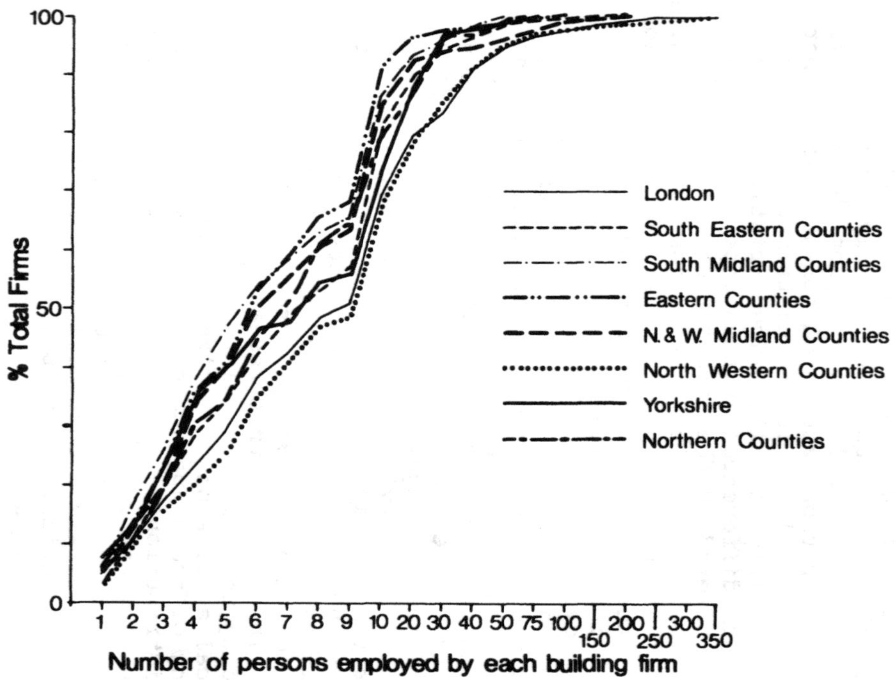
100
50
0
% Total Firms
1 2 3 4 5 6 7 8 9 10 20 30 40 50 75 100 200 300
150 250 350
Number of persons employed by each building firm
London
South Eastern Counties
South Midland Counties
Eastern Counties
N. & W. Midland Counties
North Western Counties
Yorkshire
Northern Counties

Figure 3.1 Cumulative frequency distribution curves of building firm sizes, 1851

different set of factors since demand is not so easily transferable.

On first consideration one might regard demand in the house-building industries of nineteenth-century towns to be readily moveable. After all, some 62 per cent of all houses completed in Leeds between 1886 and 1914 were back-to-back and in other northern towns the bulk of the new stock fell into particular morphological and rental categories (Beresford, 1971, p.118). However, the fact that in this industry the place of production and consumption were the same, each unit of output being tied to a specific locality, rendered every product different. Housing demand was spatially differentiated in the city and heterogeneity derived from the simple fact that each parcel of development land was not equally attractive to all builders; differences in the size and type of building projects undertaken by firms meant that their realization of locational advantages with respect to specific sites was often weakened. For instance builders' activities in the short term had an essentially punctiform spatial pattern. They tended to build within a convenient travelling distance of their own homes and yards and, if the urban frontier shifted beyond this economically viable 'field', they would generally follow. The logistics of the industry only rarely permitted these specialized house builders to execute projects contemporaneously in different parts of a large city. Moreover, once they had familiarized themselves with a particular market demand, they tended to specialize in that class of housing. On the other hand, the small, non-specialized builder had a much greater freedom of choice of site but could operate at only a small scale of output. Even within one locality the characteristics of parcels of development land could vary greatly in terms of size and shape, the attribute of extension, adaptability to specific plot morphologies, the policies of ground owners, proprietary rights, and usages of adjacent land. Thus, the uniqueness of location rendered each speculative house-building project a distinct unit of output. Given such heterogeneity, a number of important concepts may be related to variations in the house-building industry.

In particular, recent work on the size distribution of firms has shifted from theoretical investigations of the static cost curve to detailed empirical studies. These have revealed that in general the long-run average cost curve does not

have the characteristic U-shape.   Bain (1956),
perhaps the first to break with classical theory in
the mid-50s, and Wedervang (1964) show that a substantial number of industries possess J-shaped
curves in which unit costs vary only slightly with
the size of firm, or approximate the horizontal
above some minimum size or critical scale.   Firms
below such a threshold represent a substantial percentage of the total market in only a very few
industries.   Any size above the minimum is
efficient since returns to scale are then constant.
Particular size distributions and levels of concentration, rather than being independently determined, are thus a function of other factors such as
the birth of new establishments and differential
growth rates.   As yet, the paucity of econometric
studies of the birth, growth, and death of firms has
retarded the development of theory to represent the
relevant dynamic processes, and the systematic
investigation of the dynamics of firms in house
building, an industry renowned for its volatility,
could provide a good testing ground for such
theoretical studies.

Thirdly, there is scope for more research on
the relationships between the size distribution of
building firms and the performance of the industry.
A recent study of the housing system argued that
the absence of inquiries into 'how private
developers ... reach the decision which finally
determines how many houses will be provided ...
(and) how different-sized firms react, or what
influence the structure of the industry has' makes
this 'an obvious and very important direction for
work in the future.'   (Murie, Niner, and Watson,
1976, p.154).   It may be suggested, for example,
that an increase in the proportion of small firms
in a population would be accompanied by greater
mobility amongst such firms and a consequently
higher degree of responsiveness to changes in
market conditions.   A trend towards concentration
would alternatively yield greater rigidity in
behaviour.   If this were the case, then it would
not be unreasonable to hypothesize a causal
relationship between the configuration of size distributions and the degree of overspeculation in
the upswing of the building cycle.   Further, this
relationship may have been strengthened in the long
term by the operation of a learning process within
the population itself, whereby, as entrepreneurial
experience accumulated, firms learned to adjust
their ways of working to cyclical fluctuations.

Little empirical work has been pursued along

these lines, as most historical studies ignore firm size as a determinant of aggregate supply. Wilkinson and Sigsworth (1977, p.207), for example, observe that the few models of the supply of private housing starts in the U.K. '...are essentially concerned with calibrating the relationship between costs and revenues on the conventional assumption of profit maximization and take as given the constraints on output imposed by the size and structure of the building industry at any point in time. There is an implicit assumption that firms are willing and able to substitute their outputs in response to differential profit rates'. Lewis (1965, pp.136-7) did argue with respect to the metropolitan house-building boom of the late 1870s that '...excessive building was almost certain in a highly speculative industry composed of both large, well-organized firms and small-scale opportunists'. Building industries in most towns do appear to have been highly speculative, and they certainly contained both large and small firms of the types suggested, but, as Saul (1962-63) has shown, they were by no means 'almost certain' to produce excessive building, or even over-building of a more modest kind.

Some recent work on long-term studies of local housing markets promises to shed light on this neglected topic. In particular, Wilkinson and Sigsworth (1977) take explicit account of the capacity of the building industry, in addition to the factors affecting uncertainty, expectation of profit and the willingness to produce, in their study of trends in property values in the three Ridings of Yorkshire. The number of workers employed (a surrogate for the number of firms engaged in construction), the number of housing starts (lagged), and the number of bankruptcies in the building industry (lagged and unlagged) are used to show the degree of capacity utilization in building at any point in time (i.e. the ability as opposed to the willingness of the industry to expand its output). The two most important of these variables, the number of bankruptcies during the current period and previous quarter together with the number of mortgage advances made by building societies, are shown by a stepwise regression technique to account for 90 per cent of all private starts during the period 1970-74.

Similarly, a recent investigation of the German house-building industry's adjustment to demand in the nineteenth century has placed greater emphasis upon the characteristics of internal

structure than comparable studies of Britain's
building cycles. Lee (1978, p.291), observes that:
'The short-term housing market seems to have been
dominated by local idiosyncrasies, which, if anything, became even more pronounced after 1880. The
structure of the local building industry and the
quality of the entrepreneurship influenced responses
to demand, while the peculiarities of local government compounded the housing situation further.'
Moreover, he suggests that the small-scale structure
of the industry permitted a high response to demand:
'Little objective economic reason existed why the
building industry could not generally adjust output
rapidly. The vast majority of builders were small
men, operating overwhelmingly on credit, employing
little fixed capital. And, in practice, the
building industry did achieve striking changes in
output over short periods.'

Fourthly, the possibility that the two main
sectors of building activity, residential construction and contracting work, were functionally interlinked by the firms themselves has only recently
been investigated (Rodger, 1975, 1979), and such
work suggests links with studies of the relationship
between building cycles and the land-use composition
of urban development (Whitehand, 1981). The conventional practice of isolating these components
for purposes of analysis may have certain advantages,
yet it hinders a more complete understanding of
entrepreneurial decision-making in the building
industry. Potential counter-cyclical income was
certainly available in this way, for house-building
was usually expanding when non-residential building
was very low; conversely, counter-cyclical forces
would tend to depress house-building just at the
time when industrial building was attaining its
peak. While the deficiencies of data for indices
such as brick output (Shannon, 1934) push one
towards generalizations about the industry as a
whole, there are other sources which allow us to
examine the degree of functional interconnectivity
between the two sectors. Lists of tenders
published in *The Builder* and other trade journals
provide information, albeit selective, on the
involvement of individual firms in contracting, and
these can be related to details in building plan
registers of house building and other projects for
which approval was granted.

Fifthly, examination of market shares and
processes of concentration in the provincial industry suggests that different types of production
were structured on the basis of markets and sub-

markets of varying size.   Before the First World
War it appears that the majority of house and
contracting builders achieved local market shares,
(that is, within particular cities), while only a
few specialized branches in the latter sector
operated at a regional or national scale.   Further,
it may be argued that much twentieth-century concentration has taken the form of a geographical
spreading by contractors, so that they now achieve
regional shares, which are the equivalent of local
shares in the earlier period.   Beyond these general
statements, we know little about the activity spaces
of various types and sizes of builder, and how these
changed during the Victorian period.   Some large
master builders attained city-wide shares during
the speculative outburst of the 1890s, but it is
likely that the majority of firms confined their
activities to particular localities in the short
run (Crossick, 1978).   In so far as interest has
focussed on the responses of builders to growing
demands of the market, it has looked at changes in
the volume of output rather than changes in the
market spread of activity.   Why some builders
thought it worthwhile to operate on such a large
scale that they could undertake a number of substantial contracts or speculative works at the same
time is an important question for future research
(Hobhouse, 1971).
    Finally, there is need to integrate work on
production and demand in regional and local sub-
markets with research on aggregative views of
building activity, and earlier studies of long
swings in the nineteenth century.   Moreover, with
information now becoming available on the size,
organizational structure, and output of various
sectors of the industry, there is perhaps scope for
refinement of some of our judgements about the contribution of the building industry to capital
formation and the course of capital and related
fluctuations in the British economy.   Indeed, with
Feinstein's new estimates of capital formation,
1760-1860, and his earlier work for the period after
1855 (Feinstein 1978, 1972), there is now a context
within which the results of local studies may be
placed.   In particular, there is urgent need for
estimates of the relative contributions of
different kinds of building firms to the process of
capital formation, incorporating, of course, the
work of maintenance, repairs, and alterations in an
activity where stock is so large in relation to the
annual output of new products.   For example, it
should be possible, through a careful use of

statistics of incremental changes in rateable value,
annual house building, and annual totals of
planning applications for various types of work, to
calculate the proportion of a particular city's
additional capital in the form of 'new construction'
that was derived from house building on the one
hand, and public, commercial, and industrial
building on the other, and how much in each sector
was dependent on the very numerous small firms, or
on medium-sized firms, and large establishments.
Having estimated these proportions, it is then
possible to investigate the origin, endurance, and
growth of firms in the several sectors with the
advantage of knowing more clearly where the main
weights of building effort were located.

Data and method

The availability of building plan registers con-
taining listings of specifications of approved
applications for building permission, creates considerable opportunities for research in this
relatively neglected area of the industry's internal
structure (Aspinall, 1978).   The author has used
these registers to examine the ways in which the
decisions of private residential developers contributed to the building process in three late-
nineteenth century provincial cities: Nottingham,
Leeds and Sheffield.   This essay looks at the
significance of the private developer in the last of
these places and particularly at the sources and
problems in analysis that are likely to be met.
The most comprehensive information on builders and
developers in Victorian Sheffield is to be found in
the fortnightly minutes of the new works and plans
sub-committee of the highway committee in the
Borough Council, a body which came into being in
October 1864 to administer the terms of the first
code of building bye-laws.   These minutes include
the names of the owners of approved plans, a description of the property detailed on such plans (which
could refer to buildings, alterations or additions
to existing structures, or new and extended roads),
and the street in which this work was to take place.
The minutes contain a number of advantages over more
traditional sources such as deeds and registers of
leases, upon which studies of the building industry
could be based.   Most important, they provide a
measure of size, in the form of planned production,
which covers the entire range of 'owners' - the
persons who took the several initiatives to sink

their capital into building - and are consistent
throughout that population. Such comprehensiveness
is particularly important when set against the fact
that much of what we know about the structure of
the industry is drawn from the records of only a
very small number of firms, and in many cases is
economically skewed in favour of the leaders.
Similarly, Vogel (1976) comments that literature on
the American building industry '...forms an inverted
pyramid with vast quantities of books on the exalted
work of the architect at the top, fewer on the work
of the various craftsmen...in the middle and, at the
bottom, the pitiful handful on the work of the
actual builder.' While the building plan applications - like the building permits used by Warner
(1962) and Simon (1971, 1978) in their studies of
nineteenth-century Boston and Milwaukee - do not
yield the sort of information contained in business
records, they do give several facts about each of
the thousands of projects which small- and medium-
sized builders were continuously executing on and
about the frontiers of nineteenth-century cities.

During the 36-year period chosen for the
Sheffield study (1865-1900), the population was
defined as every application for residential
property and new or extended streets approved by
the sub-committee, those disapproved not being
listed. In the case of house building these
numbered 9,295 and contained a total of 53,651
houses; those for new streets exceeded 500 and
detailed a total of 993 new or part-new streets.
Upon the basis of name differentiation, a total of
4,069 persons emerged as owners of house plans, 118
of whom also owned street plans, while an additional
118 persons owned street plans only.

Very little can be said about the identity of
these 4,069 private residential developers from the
minutes, but it is likely that there were at least
three groups of people who contributed to urban
building. First, there were the speculative
builders who actually built the houses. This
category included both builders who built
speculatively for sale as rapidly as market conditions would allow, and builders who acted as landlords and erected houses as a speculative investment.
Secondly, there were the speculative developers.
These were people who formulated the plans,
assembled resources, managed the transformation of
these resources into new housing, and sold that
product, as a specialist in the entrepreneurial
side of housing creation but not necessarily as a
house builder. Finally, there were the non-

specialists who arranged for the erection of one or more houses by entering into contracts with builders or craftsmen. These would then be let, sold, or occupied personally. Amongst those in the first category, the builders themselves, not all would be 'house builders' in the sense that we understand the term, that is, firms which undertook responsibility for the construction of the entire building, either by employing only their own workers or by sub-contracting to other craftsmen. Rather, there must have been great variety of industrial organization, in terms of size, specialization, and continuity. Dyos (1968, p.654) suggests that the main body of house builders in Victorian London - the 'thousands of builders who kept apart from the main professional body and have preserved their anonymity ever since' - would have included both 'the small jobbing builders who broke off now and then from their repairs and alterations to put up a house or two' and 'the large and the small speculative builders who kept in close pursuit of the suburban frontier and, indeed, scarcely did anything else'. In reality, we have little direct evidence as to whom we are describing, but a large number of those at work in provincial cities must have engaged in house building as a part-time pursuit as 50 per cent of 'owners' in late-Victorian Sheffield erected less than four houses altogether (Aspinall, 1977). The proportion of building owners which belonged to this group of speculative house builders is difficult to establish since the names of plan depositors were not recorded in the minutes. However, using registers for London and working on the assumption that property was built on specula-tion in those cases where there was a coincidence of name for depositor and owner, Dyos (1968) calculated that some 90 per cent of house property in south London was so built and the proportion speculatively built in London suburbs in the 1880s was put as high as 99 per cent by a leading real-estate developer (Select Committee on town holdings, 1887). It is perhaps not unreasonable to suggest that in rapidly expanding northern industrial towns the figure would not be radically different.

Not only was the array of building firm types in the Victorian city extremely complex, but the line of demarcation between speculative builder and developer was also a shadowy one. In London the developer was the link between the capital market and the building process, while in the industrial towns of the north he either emerges somewhere at the top of the hierarchy of speculative builders or

86

as a small-scale estate owner. Some of these
actors can be identified in the names of street
plan owners, the registers also revealing their
involvement, if any, in house building. They
included, for example, Thomas Steade, owner of
plans for 309 houses and 25 streets; J. Hayhurst,
owner of plans for 83 houses and 25 streets;
F.U. Laycock, owner of plans for 103 houses and 15
streets; and perhaps many of the 17 other persons
who owned plans for 16 or more houses and 4 or more
streets in Sheffield (Table 3.1). There were also
the landowners who took on the role of developers,
notably in Sheffield, the Duke of Norfolk, owner of
plans for 83 streets; M.G. Burgoyne, owner of plans
for 52 streets; and Earl Fitswilliam, owner of
plans for 28 streets. Clearly, the entrepreneurial
function of 'speculative builder' and 'speculative
developer' must have increasingly merged as the
scale of house-building projects increased. Lastly,
many of the owners of plans of one or a few houses
were likely to have been intending owner-occupiers
or small-scale investors whose contribution to the
building process was an isolated or short-lived
event.

In addition to comprehensiveness, data on size
of 'firms' detailed in the Sheffield minutes -
planned output expressed in its most disaggregated
or unit form - also has certain advantages over
other measures of production. No statistics of
capital expenditure were systematically collected
and what data are available suggest that the
industry was very lightly capitalized as far as
fixed assets were concerned. Although some contracting firms specializing in large-scale housing-
building projects were more heavily capitalized, the
capital requirements of jobbing builders, firms
working mainly on the repair and maintenance of
houses, and the vast number of small house-building
firms would be tiny. Likewise, no information is
available on the number of persons employed. Even
if it were, such a measure would have little
superiority over output. The level of employment
in house-building firms was subject to marked
fluctuations according to the state of trade.
While many of the smaller firms appear to have kept
a permanent workforce, the large contractors
engaged most of their site labour on a purely *ad hoc*
basis and retained only a small number in permanent
employment. These characteristics of labour and
capital suggest that concepts of 'actual capacity'
and 'full capacity' size are of little
relevance, and that while no single indicator

Table 3.1: The ownership of street plans in Sheffield, 1865-1900

		Number of Streets															
		1	2	3	4	5	6	7	8	9	10	11	12	13	14	15+	
Number of Houses	0	38	24	18	17	5	5	1	1	2	-	-	1	1	-	5*	118
	1	3	-	1	1	1	1	2	-	-	-	-	-	-	-	-	9
	2-3	2	2	2	-	-	1	-	1	1	-	-	-	-	-	2**	11
	4-7	4	1	1	-	1	-	-	-	-	1	1	-	-	-	1***	10
	8-15	5	5	-	1	1	1	-	-	-	-	-	1	-	-	-	14
	16-31	7	2	1	1	1	1	1	-	-	-	-	1	-	-	-	15
	32-63	9	2	3	2	-	-	-	1	-	-	-	-	-	-	-	17
	74-127	8	4	1	2	-	-	-	-	-	-	-	1	-	-	2****	18
	128+	13	3	1	2	2	-	-	-	-	1	-	-	-	1	1*****	24
		89	43	28	26	11	9	4	3	3	2	1	4	1	1	11	

* 15; 16; 24; 26; 52

** 15; 28

*** 83

**** 15; 25

***** 25

of size is satisfying from all points of view, the best would be output itself. Unlike fixed capital, it has no age dimension and its independence of value makes it more suitable for inter-temporal comparisons. The only obvious difficulty lies in inter-sectoral comparisons, especially between house building and larger scale projects.

Nevertheless, there is one serious drawback if these minutes are used to define size structure, namely, that they make no allowance for variations in the sizes of houses. Clearly, the seriousness of this omission depends largely upon the degree of homogeneity amongst builders' productions. Additions to the housing stock of the northern industrial towns of late-Victorian England show little differentiation. Sheffield had an over-whelmingly working-class and industrial character, the housing market demanding, according to one writer, '...the maximum number of the cheapest possible cottages, crowded as closely together as possible around places of employment ... reconciling the single-family house with a compact, high-density community' (Olsen, 1973, p.353). In support of this view, an analysis of gross estimated rentals in rate books shows that by 1887 68 per cent of the 65,500 houses in the Borough had weekly rents of 3s 10d and less per week, while in the more heavily industrial colony of Brightside Bierlow this rose to 73 per cent.

While the Sheffield minutes, like most registers for other towns, list details of projects which were approved, they omit information about construction itself. Thus, some estimate is required of the overstatement of the amount of activity detailed in the plans and the time lag between planning approval and actual construction. Several research workers have calculated from such dual sets of figures as they had, that between 10 per cent and 15 per cent of the annual building plans failed to be executed, the resultant series being lagged by six months or the time which they considered it took to build a house (Weber, 1955; Lewis 1965). Obviously no procedure can be suited to all data, but it has to be borne in mind that the percentage of redundancies did vary quite markedly from one town to another and even within individual towns from year to year. In Sheffield almost 30 per cent of the total houses in plans approved between 1865 and 1900 were never built, while in Leeds the average for the period 1886-1914 was 17 per cent. In fact, in the analysis of Sheffield's house-building industry, no attempt has

been made either to convert approvals into erections
or to lag the series since interest lies as much in
the response of firms to changes in market conditions at the planning stage of the building process,
as in statistics of actual output, and because such
methods of statistical reduction cannot be extended
to the totals for individual builders, since there
is no way of knowing where in the size array this
over-anticipation was most rife.

Finally, a more general point must be noted
about the data. The use of data compiled for
specific areas, primarily of administrative rather
than economic significance, may result in an
erroneous trend in building activity or the misesti-
mation of the sizes of firms. This could arise
from changes in the boundaries of the areas to which
the data relate over the period under consideration,
or from circumstances in which urban areas were
seriously underbounded. However, the data in the
Sheffield minutes do not suffer from such inconsistencies since the function of the sub-committee did
not change in the period, and all the plans approved
by it related to the area of the Borough, the extent
of which remained fixed until 1901 and contained all
the building that took place in the city itself.

The housebuilding industry in Sheffield

The remainder of this essay discusses selected
aspects of the author's investigation of the size
structure of Sheffield's late Victorian house-
building industry. Further details of this study
may be found in Aspinall (1977), as only the more
general conclusions about the size distribution of
building firms can be reported here to illustrate
the sorts of analysis that may be undertaken.

The size of building projects

The extent to which cyclical fluctuations affected
the size structure of building projects can be
elicited from the 9,295 building specifications.
Figure 3.2a clearly shows that the configuration of
this distribution is positively skewed. The overall importance of small-scale projects is very
apparent, 20.9 per cent of all applications throughout the entire period relating to one house only
and a further 20.2 per cent to just two houses. In
all, almost two-thirds of applications refer to
projects of less than five houses. Thereafter, the
size groups employed show a fairly regular decline,

despite a secondary modal peaking in the 20-29 class. The only comparative statistics relate to building project sizes in Leicester during the period 1850-1900, 70 per cent of which involved less than five houses. In 1870 105 applications were submitted to the Council for permission to build, 79 (75.2 per cent) of which were of this size and only one (0.95 per cent)was for more than 20 houses (Potts 1970; Pritchard, 1976).

If we look at the size structure of Sheffield's projects at discrete three-yearly time intervals through the use of cumulative frequency distribution curves, (Figure 3.2b) several marked trends emerge. As more large projects are undertaken, the upper right-hand tails of the curves become increasingly distended until they tend towards a straight line particularly from the late 1880s. Except for the first curve (1865-7), the smallest distribution of building project sizes, the other curves are tightly clustered until that for 1887-91 is encountered. By 1898-1900 there is a clear shift away from the main trend, marking the burst of speculative building activity at the end of the Victorian period.

Building firms: size distribution

Let us now move from the projects to the building firms themselves, making the reasonable assumption that the number of planned houses intended to be erected by builders bears some significant relationship to the overall size of the building firm (Dyos, 1968). A total of 4,069 different building owners are named in the specifications during the period 1865-1900, 23.2 per cent of whom submitted plans for one house only, and 18.8 per cent for just two houses. Half the builders put up no more than three houses, three-quarters no more than 8 houses, and 87.0 per cent under 20 houses. Really large builders, those erecting 100 houses or more, account for a mere 2.4 per cent of the first (Table 3.2). With a few exceptions these firms were overwhelmingly proprietorially managed. Scarcely a dozen house-building enterprises formed themselves into joint-stock companies during these years, and James Longden & Co. Ltd., the town's largest contractor, did not take the step until the early 1900s. The J-shaped long-run average cost curve elaborated by Wedervang (1964) seems to fit these empirical findings of marked positive skewness very closely. Its appeal lies in the fact that large numbers of firms of the minimum size (those building one or two houses) were present in the industry, that they

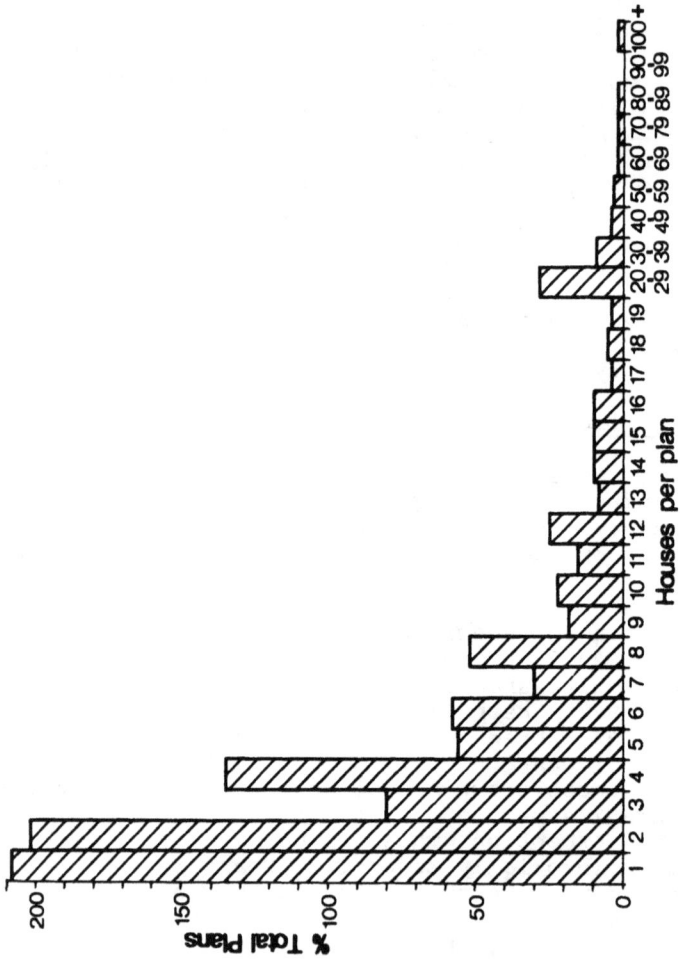

Figure 3.2a Histogram of building project sizes, 1865-1900

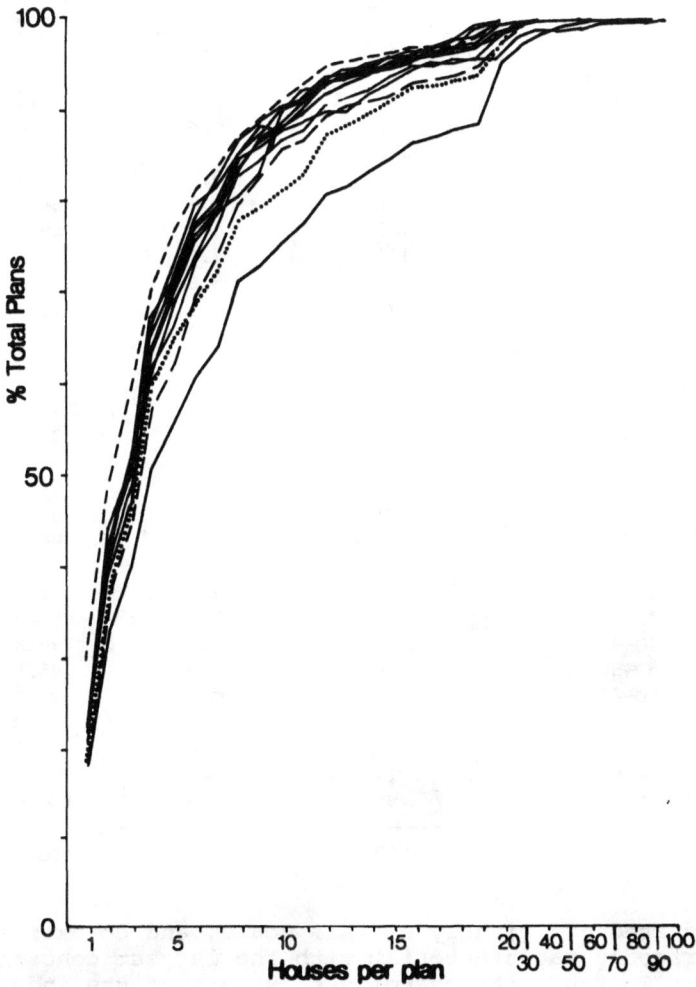

Figure 3.2b   Three year cumulative frequency distribution
curves of building project sizes, 1865-1900

Table 3.2: House Builders in Sheffield, 1865-1900

Number of Firms

Size of Business (no. of houses in plans)	absolute frequency	relative frequency (%)	cumulative frequency (%)
1	942	23.2	23.2
2	765	18.8	42.0
3	336	8.3	50.2
4	402	9.9	60.1
5	176	4.3	64.4
6	183	4.5	68.9
7	103	2.5	71.4
8	147	3.6	75.1
9	76	1.9	76.9
10	76	1.9	78.8
11	46	1.1	79.9
12	67	1.6	81.6
13	42	1.0	82.6
14	43	1.1	83.7
15	32	0.8	84.4
16	39	1.0	85.4
17	21	0.5	85.9
18	26	0.6	86.6
19	18	0.4	87.0
20-29	175	4.3	91.3
30-39	76	1.9	93.2
40-49	54	1.3	94.5
50-59	39	1.0	95.5
60-69	31	0.8	96.2
70-79	25	0.6	96.8
80-89	17	0.4	97.2
90-99	14	0.3	97.6
100+	98	2.4	100.0
Totals	4,069	100.0	100.0

operated efficiently at this scale, and co-existed perfectly satisfactorily with the largest concerns.

To study the changing structure of the industry in more detail, one must first define the span and shape of the size distribution. If we compare the total number of firms active from year to year with the building cycle for Sheffield, it can be seen (Figure 3.3) that this number varies annually, these changes bearing a reasonably close relation-ship to fluctuations in the building cycle. In the trough of 1865 they number 173 but rapidly increase on the upturn to average 246 per annum in the four

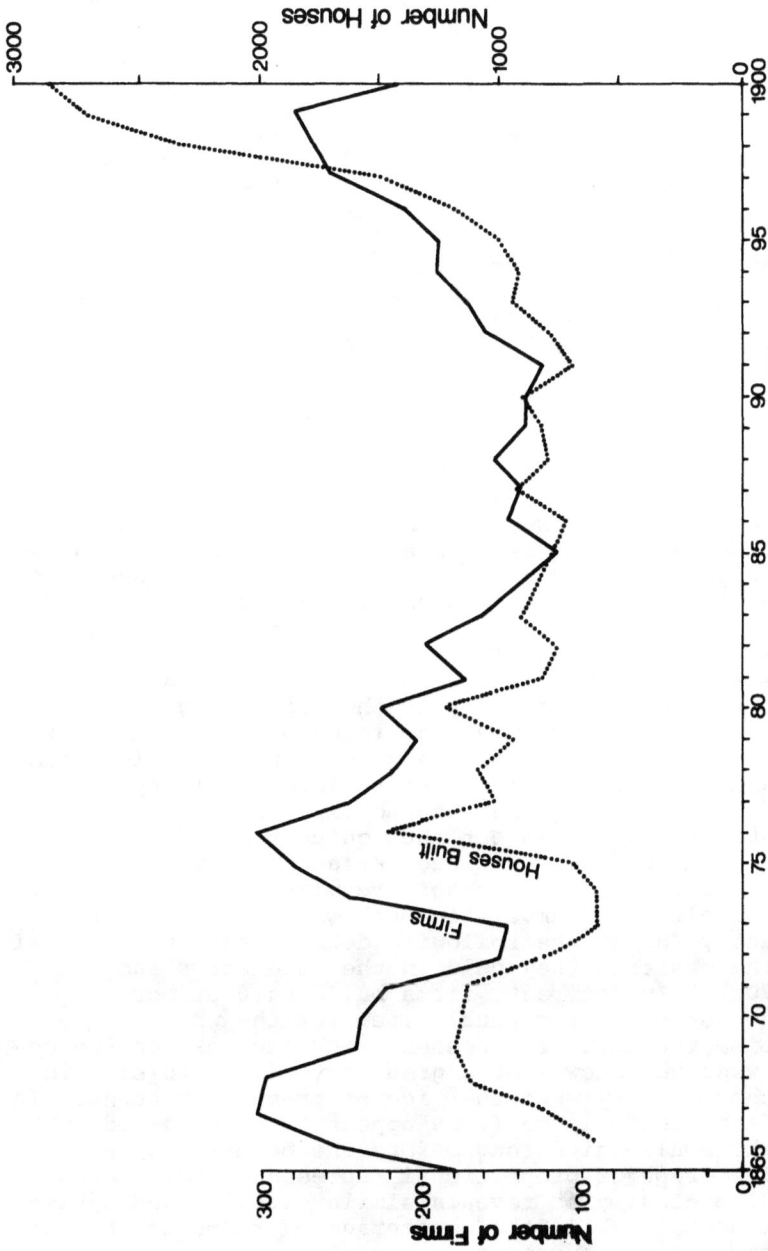

Figure 3.3 The housebuilding industry in Sheffield

years comprising the crest of the boom of 1868-71. In the first two years of the trough they fall back to 151 and 145 but again rapidly recover on the up-turn to reach an all-time maximum of 298 in 1876, the peak year for house completions prior to 1897. Productions and the number of firms fell away again in the next three years but the latter rise to 220 in 1880, the second peak year for house completions prior to 1897. From 1881, however, annual house completions are consistently low and slowly slide to the bottom of the trough in 1891, the numbers of firms during these slump years falling to minima of 114 and 124 in 1885 and 1891, respectively, and averaging 151.4 per annum (1881-94). On the upturn the number of firms increases each year from 159 in 1892 to 188 in 1895, the first boom year, then soars to 279 in 1899, the peak year for the submission of building plans, the annual average for the six years being 235.2. Even more dramatic is the rise in the number of house completions, from an average of 836 in the depression years 1881-94 to an average of 1,934 in the succeeding boom years.

Within this varying number of firms, the absolute and proportional representation of each size class also changes in a number of significant ways (Figures 3.4 and 3.5). Firms building 1 house, 2 or 3, and 4-7 houses show an overall but irregular decline throughout the whole period. Each of these size classes contain over 70 firms in 1867, the start of an upswing in the long cycle, but all three decline markedly in the early stages of the 1868-71 boom to around 50 in number. At the bottom of the trough in 1873 each contains 40 or less firms. While the revival in house-building activity is delayed till 1876, all three experience a peak in numbers in 1875/6 but then quickly fall away, even though building continues at a high level. This decline continues in an irregular manner until 1889 and although numbers in each group rise significantly during the following decade, they never regain the position they held in the late 1860s and mid-70s. In contrast, firms building 8 or more houses show a gradual rise over the whole period. Like the small businesses, they increase on the up-swing but show a much greater tendency to stay in the industry when the tide of prosperity turns. In fact, these large firms appear to have forced out the small units long before the boom was spent. The graph of proportional representation of each size class also reveals similar secular and cyclical trends. While the percentage of firms in the three smallest size groups varies widely from year to

96

year, the relatively greater rise in the numbers of large firms at the expense of the small is shown in the general decline throughout the whole period. The substantial amount of fluctuation from year to year amongst those firms building 1 to 7 houses appears to decrease as one moves up the size array, the largest size group of 32+ houses showing a high degree of stability until the upswing in the late 1890s.

In explaining the rise of large building firms in the size structure of the house-building industry in Victorian London, Dyos (1968) suggests a demand-related model of firm-size growth, the expanding demand for suburban houses leading directly to a change in the structure of the industry which supplied them. Similarly, in late Victorian Sheffield, the graphs suggest that it was prosperity which gave rise to the fully developed master builders, invoking a positive relationship between firm size and demand. This association was particarly close for the three largest size groups during the great boom of 1896-1900, as they increasingly displaced the small units in terms of proportional importance. To measure the degree of agreement between the relative importance of large firms and the size demand, product moment correlation coefficients were calculated between the percentage of building firms in each of the three largest size groups and the total number of planned houses per annum between 1865 and 1900. A very good relationship is obtained for firms building 32 or more houses (+0.84), weakening to +0.26 in the size groups 16-31 and 8-15 houses respectively. These observations are reinforced by other characteristics of the graphs. Given certain threshold sizes of demand, bigger firms seem to come into the industry. The boom of 1868-71 appears to have permanently brought in the 8-15 and 16-31 groups, after which their percentage importance rises by fairly constant increments. The curve for 32+ houses is perhaps the most interesting so far as arguments concerning the emergence of the large-scale, fully-fledged master builders are concerned. The percentage importance of these firms scarcely varied until 1896, oscillating around the 3 per cent level, but in that year, marking a burst of speculative building activity, suddenly increases to over 6 per cent, the peak of 11 per cent being achieved in 1899 on the crest of the building boom.

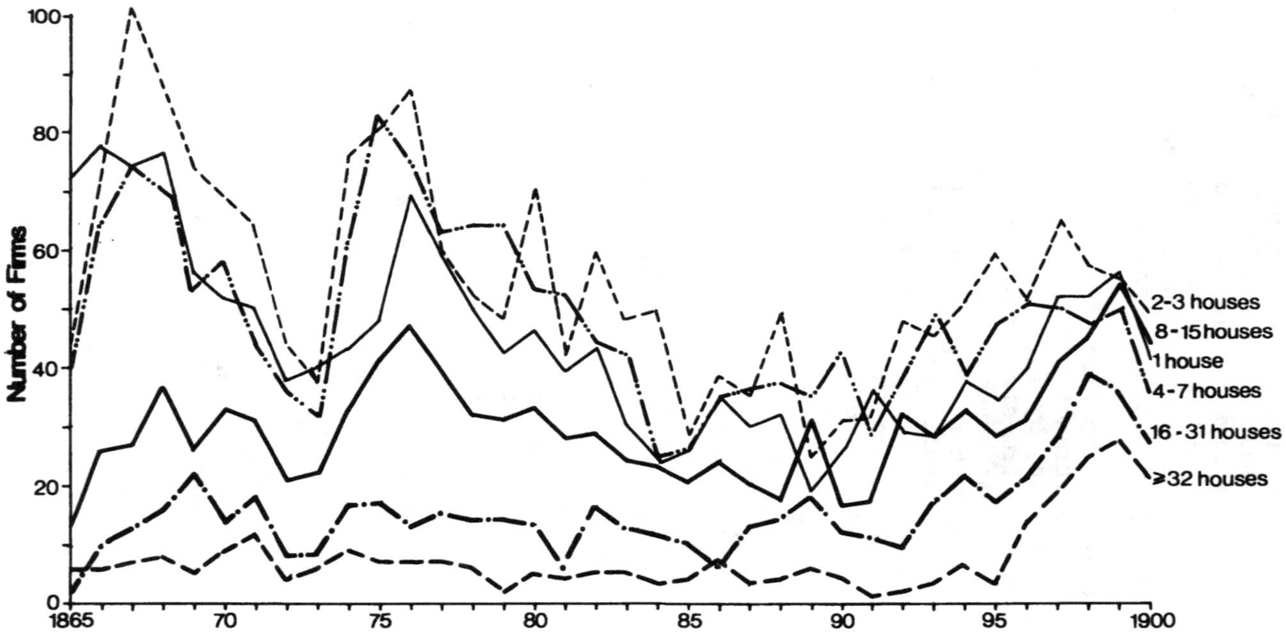
Number of Firms
100
80
60
40
20
0
1865
70
75
80
85
90
95
1900
2-3 houses
8-15 houses
1 house
4-7 houses
16-31 houses
≥32 houses

Figure 3.4 Numbers of building firms in different size groups, 1865-1900

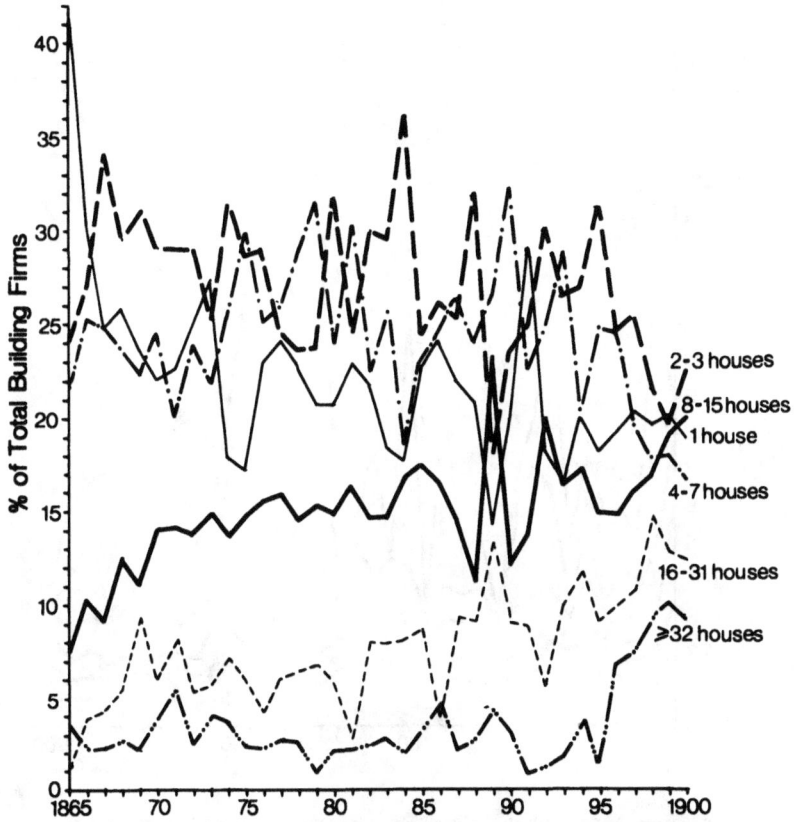

Figure 3.5 Percentage of building firms in different
size groups, 1865-1900

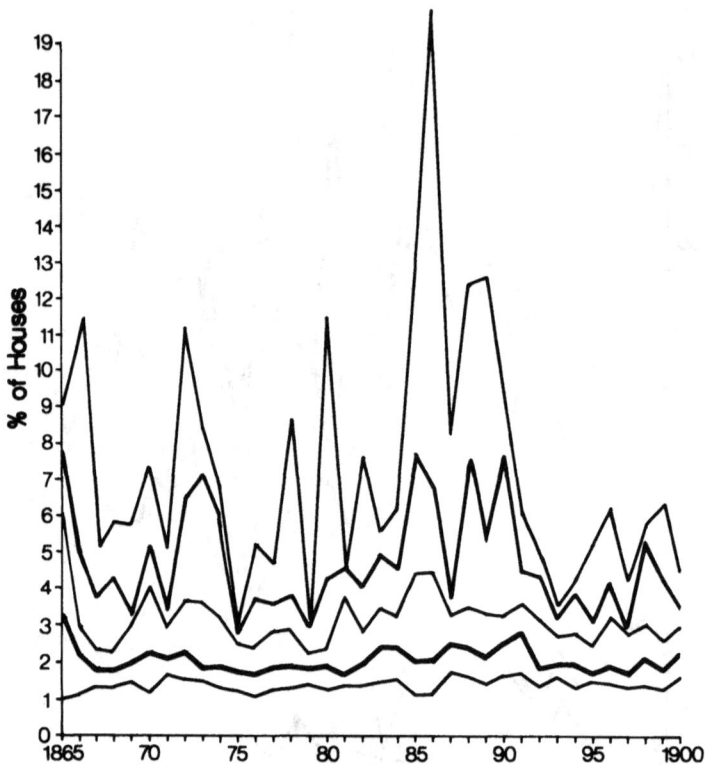

Figure 3.6a Rank order of the 1st., 2nd., 4th., 8th.,
and 16th. building firms per annum, 1865-1900 :
relative frequency of productions

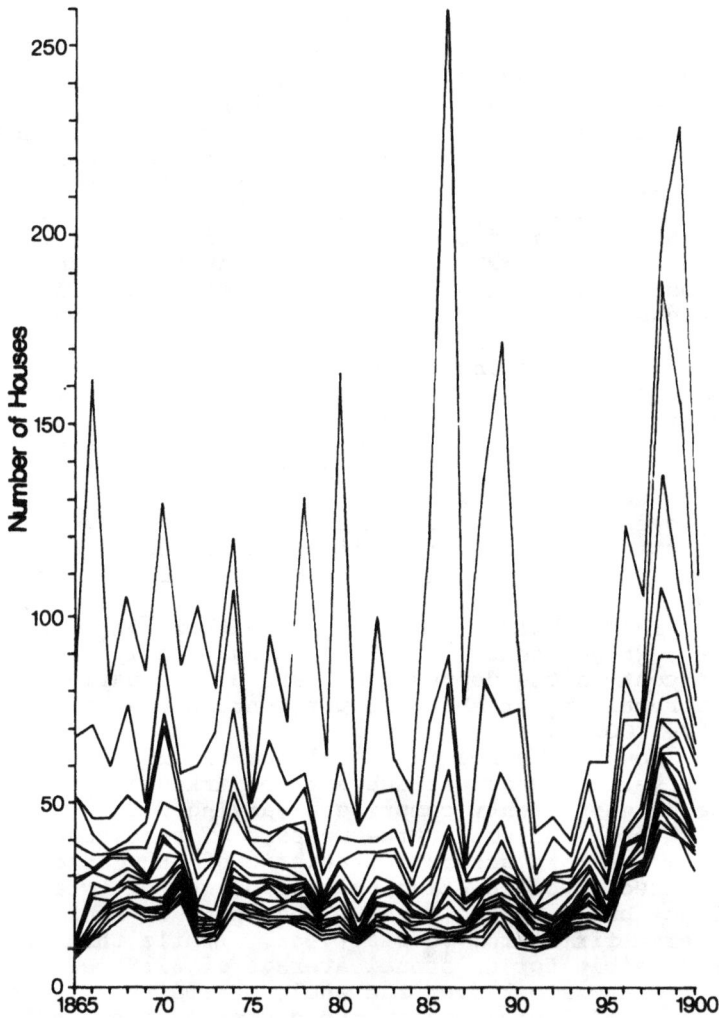

Figure 3.6b   Rank order of the 20 largest building firms
per annum, 1865-1900 : absolute frequency of productions

Building Firms: Size Concentration

We can also analyse variations in the size distribution of firms in terms of the total housing stock.
The number of houses built by the 20 largest house-
building firms in Sheffield fluctuated widely in the
1860s, '70s, and '80s, while in the depression years
of 1891-5 most of the large building firms went
under altogether (Figure 3.6b). For three of the
slump years (1891-3) the first ranking builder put
up less than 50 houses, but in the last five years
of the Victorian period (the great building boom),
the number soared to a peak of 229 in 1899. In
both 1898 and 1899 the twentieth ranked builder put
up more houses than the first ranked builder in the
depression year of 1893. Moreover, in the boom
years of the 1890s the increase in the scale of
building operations permeated virtually the whole
rank order, unlike the 1870s and '80s when only a
few freak speculators put up houses in such large
numbers. Although the appearance of these really
big firms in the late-Victorian boom is prominant
when the unit of size, the number of houses, is
employed as an absolute measure, the relative contribution of these firms in terms of the percentage
of all houses erected in any one year is not as
significant (Figure 3.6a). For example, the first
five ranked firms built 33 per cent of all houses in
1865, 25 per cent in 1866, falling to only 15 or 16
per cent in the depression years of the early '90s
and remaining at about 18 per cent in the period
1887-1900.

However, since the total number of firms active
from year to year was subject to marked fluctuations,
the largest twenty comprise a varying percentage of
this total and not a true measure of size concentration. A more accurate statistic of the latter is
obtained by calculating the proportion of total
houses built by the first 20 per cent of rank
ordered firms from year to year. While these were
responsible for an annual average of 62.9 per cent
of the new stock over the whole period the figure
fell to 57.5 per cent in the slump years of 1891-5
and rose to 67.1 per cent in the succeeding five
boom years. Thus, it is chiefly these medium and
large firms which took a disproportionate share of
the total number of houses built.

Conclusion

There is not space to discuss the internal dynamics
of the building industry nor the detailed reasons
for structural change in this essay, but some brief
conclusions may be offered.   The case study
(Aspinall, 1977) would seem to re-affirm the
importance in speculative house building of the
small and short-lived building firm (Dyos, 1961,
1968) which dominated the industry in Sheffield until the 1890s.   While the contribution of these
firms to the total output of houses was notable, it
can be suggested that their primary significance
resided in their function as a seed-bed in which
larger firms and new enterprise originated.   In the
later part of the century the rapid growth of the
housing market clearly favoured the supersession of
these firms by the more comprehensively organized
master builders, working on a large scale, although
few such Sheffield builders maintained their
position for as long as Edward Yates, the South
London speculative builder who completed about
2,500 houses between 1867 and 1895.   A structure of
this kind, composed of a great many small firms
with high rates of entry and exit from year to year,
and an increasing number of large ones, all of
which had ready access to the required capital and
the ability to avoid the outlay of a sizeable proportion of this on their building sites by virtue
of the leasehold system, may well have contributed
to the chronic over-building in Sheffield during
the nineteenth century.   A succession of informed
contemporary views focussed on this aspect of the
market.   As early as 1833, John Parker, M.P., a
member of the Select Committee on Public Works,
informed his colleagues that '...in Sheffield, the
average of the comfort of the lower classes is above
that of most other places; we have not yet got into
the abominable way of cellars, or of many families
living in the same house'.   The average dwelling
was the single-family 3- to 4-room cottage and the
average number of persons per house less than five.
A decade later Holland`(1843) ascribed what he saw
as the over-production of houses in Sheffield to
'the petty capitalist...desirous of realizing a
handsome per centage...' as much as to the landowner,' naturally anxious to appropriate his land
to building purposes', and the 'pennyless
speculative builder' (Olsen, 1973).   Haywood and
Lee (1848) also found the supply of houses
perfectly satisfactory: 'Notwithstanding all the

evils we have seen during our inspection of the town, we do not hesitate to say, generally, that the construction of the houses occupied by the working classes in Sheffield is better, and the rental more moderate, than in almost any other town in the kingdom'. Finally, in January 1907, the *Sheffield Independent* was referring to the unattractiveness of building in Sheffield which was 'undeniably overbuilt'. While a market situation of this kind could be symptomatic of many circumstances, the basis for the peculiar susceptibility of the speculative house-building industry to operate within a wide margin of error still requires much closer scrutiny.

References

Aspinall, P.J. (1977) *The size structure of the housebuilding industry in Victorian Sheffield*, Working paper 49, Centre for Urban and Regional Studies, University of Birmingham

Aspinall, P.J. (1978) *Building applications and the building industry in nineteenth century towns: the scope for statistical analysis*, Research memorandum 65, Centre for Urban and Regional Studies, University of Birmingham

Bain, J.S. (1956) *Barriers to new competition: their character and consequences in manufacturing industries,* Harvard University series on competition in American industry, 3, Cambridge, Mass.

Beresford,M. W. (1971) 'The back-to-back house in Leeds, 1787-1937' in S.D. Chapman (ed.) *The history of working-class housing: a symposium,* David and Charles, Newton Abbot, pp.93-132

Committee of Enquiry on Small Firms (1971) Merrett Cyriax Associates, 'Dynamics of small firms', research report no. 12, H.M.S.O.

Crossick, G. (1978) *An artisan élite in Victorian society: Kentish London, 1840-1880,* Croom Helm, London

Daunton, M.J. (1977) *Coal metropolis, Cardiff 1870-1914,* Leicester University Press, Leicester

Dyos, H.J. (1961) *Victorian suburb: a study of the growth of Camberwell,* Leicester University Press, Leicester

Dyos, H.J. (1968) 'The speculative builders and developers of Victorian London', *Victorian Studies,* 11, pp.641-90

Feinstein, C.H. (1972) *National income, expenditure and output of the United Kindon, 1855-1965,* Cambridge University Press, Cambridge

Feinstein, C.H. (1978) 'Capital Formation in Great Britain', in P. Mathias and M.M. Poston (eds.) *The Cambridge Economic History of Europe,* vol. VII, Cambridge University Press, Cambridge

Haywood, J. and Lee, W. (1848) *Report on the sanatory condition of the borough of Sheffield,* Sheffield

Hobhouse, H. (1971) *Thomas Cubitt: master builder,* Macmillan, London

Holland, G.C. (1843) *The vital statistics of Sheffield,* Sheffield

Imray, J.M. (1966) 'The Mercers' Company and East London, 1750-1850', *East London Papers,* 9, pp.3-25

Lee, J.J. (1978) 'Aspects of urbanization and economic development in Germany, 1815-1914' in P. Abrams and E.A. Wrigley (eds.) *Towns in societies: essays in economic history and historical sociology,* Cambridge University Press, Cambridge, pp.279-93

Lewis, J.P. (1965) *Building cycles and Britain's growth,* Macmillan, London

Murie, A., Niner, P., and Watson, C. (1976) *Housing policy and the housing system,* Allen and Unwin, London

Olsen, D.J. (1973) 'House upon House' in H.J. Dyos and M. Wolff (eds.) *The Victorian city; images and realities,* Routledge and Kegan Paul, London, pp.333-357

Potts, G. (1970) 'The Creation of Highfields, Leicester', Unpublished paper presented at the annual Conference of the Urban History Group, Birmingham

Pritchard, R.M. (1976) *Housing and the spatial structure of the city: residential mobility and the housing market in an English city since the Industrial Revoluion,* Cambridge University Press, Cambridge

Rodger, R.G. (1975) 'Scottish urban house-building 1870-1914', Unpublished Ph.D. thesis, University of Edinburgh

Rodger, R.G. (1979) 'Speculative builders and the structure of the Scottish building industry, 1860-1914', *Business History,* 21, pp.226-246

Saul, S.B. (1962-63), 'House-building in England, 1890-1914', *Economic History Review,* 2nd series, 15, pp.119-37

Select Committee on Town Holdings (1887, 1888), Evidence of E. Tewson, G. Simpson

Shannon, H.A. (1934) 'Bricks - a trade index, 1785-1849', *Economica,* New series I, pp.300-18

Sheppard, F. (1971) *London 1808-70: the infernal wen,* Secker and Warburg, London

Simon, R.D. (1971) 'The expansion of an industrial city: Milwaukee, 1880-1910', Unpublished Ph.D. thesis, University of Wisconsin

Simon, R.D. (1978) 'The city-building process: housing and services in new Milwaukee neighbourhoods, 1880-1910', *Transactions of the American Philosophical Society,* 68, part 5

Sutcliffe, A.R. (1972) 'Working-class housing in nineteenth century Britain: a review of recent research', *Bulletin of the Society for the Study of Labour History,* 24, pp.40-51

Vogel, R.M. (1976) 'Building in the Age of Steam', in C.E. Peterson (ed.), *Building early America: contributions*

105

Chapter 4

THE GROWTH OF PUBLIC INTERVENTION IN THE BRITISH URBAN ENVIRONMENT DURING THE NINETEENTH CENTURY: A STRUCTURAL APPROACH

Anthony Sutcliffe

Introduction

It is a commonplace that an important extension and intensification of the role of public authorities in moulding the British urban environment occurred during the nineteenth century. In 1800, even in the largest towns, manipulation of the environment was conceded almost entirely to the free play of the market. By 1914, so far had public intervention developed, both in the (positive) provision of elements of the urban infrastructure and in the (negative) restriction of the free use of private land and buildings, that it had come to be generally agreed that the evolution of entire urban areas could usefully be directed by a public programme of development, or plan.

Ever since the early 1950s, when Ashworth's pioneering survey appeared (Ashworth, 1954), this massive growth of public activity has fascinated historians. However, familiarity with the phenomenon has tended to breed contempt for the fundamental question of why it ever occurred at all. Studies of particular aspects of intervention, such as public-health policy or municipal house-building, rarely grasp this general issue. Municipal histories certainly cover the full range of policy, but usually fail to set it in the contexts of the economic, social and physical evolution of the town, and of national policy developments. On the other hand, national studies of social policy, in the 'rise-of-the-Welfare-State' tradition, assume too readily that urban developments conform to a general pattern of progress towards social reform. Rarely do we fully appreciate (though see Falkus, 1977) the extraordinary ambition of specifically urban modes of intervention in economic and social processes, during a century in which the informed public was

generally unsympathetic to administrative limitations of individual freedom. For instance, why
were most urban tramways in public ownership by 1914,
while most railways and canals were still private?
Why was part of the demand for workers' housing
being met by municipal enterprise, sometimes with
the help of more or less concealed subsidies, while
the production of food and clothing remained
entirely in private hands, and unsubsidised? Why
was the use of an urban site hedged about with
controls which almost certainly reduced the income
which it could generate, while the owner of a parcel
or rural land could dispose of it virtually as he
wished?

The most obvious explanation of these questions
is that there must be something unusual about towns.
If we want to understand the growth of public intervention in the development of the urban environment
at a time when other areas of production were left
almost entirely to private enterprise, we need to
look closely at the free-market processes involved
in the physical development of towns. How far were
these processes alone capable of producing an
acceptable urban environment? Did their capacity to
do so vary over time? Did they cater equally well
(or equally badly) for the environmental needs of
all members of the urban community, or were some
town-dwellers worse served than others?

The origins of public intervention

First, it will greatly help if we recognize that the
production of an acceptable urban environment has
always required a degree of public intervention.
Nineteenth-century urban authority, far from
initiating intervention in response to 'new' problems, inherited a basis of powers and practices
from the pre-industrial period. This continuity
tends to be obscured by the creation of numerous
special urban authorities (Improvement Commissions)
between the mid-eighteenth and early-nineteenth
centuries (Webb, 1922), and the apparent inactivity
of most of the unreformed corporations (Chinnery,
1974). However, the main development during the
period of early industrialization was simply that a
large number of communities were reaching such a
size and degree of structural complexity that concerted public intervention became necessary. In
effect, a pattern of intervention which had been
perfected in the larger towns, particularly London,
since the middle ages, was diffused and grafted onto

the rudimentary functions of environmental control which since Tudor times had been entrusted to the parishes (Webb, 1913).

If we had visited London or one of the big provincial towns in the early 1800s, we would have found public authorities already intervening in four major aspects of the urban environment:-

    (i)   the provision and maintenance of thoroughfares

   (ii)   the provision and maintenance of drainage

  (iii)   the control of building to reduce the danger of conflagrations

   (iv)   the control of smoke and other noxious emanations.

If we had stopped to consider why this administrative activity was necessary, we would almost certainly have concluded, perhaps aided by developments in utilitarian social theory at that time, that the agglomeration of individuals and properties in a town, though economically and socially beneficial, required a degree of cooperation and restraint largely unnecessary in rural areas. One might reasonably assume that the owner of an urban site would seek to provide access to it and to drain it, but he could achieve these ends only by establishing costly easements over or under neighbouring properties. With his neighbours trying to do the same, the cumulative cost of these separate negotiations would rise out of all proportion to the benefits secured, discouraging some or all owners from making effective arrangements. Ultimately the deficiencies of access to, and drainage of, individual properties would detract from the general good of all owners and occupiers. In consequence, owners would normally agree to the creation of unified systems of thoroughfares and drainage, either through co-operative action or toleration of the action of a superior authority.

Conflagrations and smoke raised a slightly different issue. In pursuing his own advantage, the owner or occupier might engage in activities damaging to his neighbours without himself suffering sufficiently to feel the need to desist from them. For instance, an owner might make a perfectly rational decision to build a store of cheap, inflammable materials, consciously incurring a higher insurance premium. If his store were in open country such a decision would concern only himself, but in a town his store, in catching fire, might bring about the destruction of an entire district, perhaps composed of uninsured buildings. Similarly, a simple flue could remove harmful or unpleasant

fumes from a workshop efficiently, only to poison
the neighbours.   In such cases, the pursuit of
individual self-interest, however rational, could
not be guaranteed to produce a result acceptable to
society.   Public regulation was thus both necessary
and justifiable even within an individualistic
ideology.

The development of public intervention in the nineteenth century

Four modes of explanation

We can now go on to consider why public intervention
in the urban environment expanded from this basis
during the nineteenth century.   Four main modes of
explanation are open to us, which for convenience we
may term adminstrative, ideological, technological
and structural explanations.

The *administrative* explanation depends on the
assumption that a bureaucracy can function to some
degree as a distinct system within society.   It was
popular in the 1950s and 1960s during the debate on
'the nineteenth-century revolution in government',
when there was much talk of the initiating role of
the professional in urban government, and of dynamic
relationships between central and local administra-
tions (Cromwell, 1966).   In recent years it has
been less influential.   Its main weakness lies in
the assumption that there is little relationship
between administration on the one hand and the needs
of society as perceived by the public or their
elected representatives on the other.   Any extension
of administration normally leads, however, to higher
taxation or greater restriction of the freedom of
the individual, both of which have sooner or later
to be defended in the political arena or in the
courts (McCord, 1978).   The failure of the 1848
Public Health Act was merely the most notorious of
numerous episodes, both national and local, in which
the administrative dynamic dashed itself to pieces
against an implacable public opinion.

If the state of public opinion is to be recognized as an important influence, there would appear
to be some scope for an *ideological* explanation of
public intervention in the urban environment.
During the nineteenth century sustained intellectual
effort was devoted, at various levels of abstraction,
to consideration of the nature of individual
happiness and the well-being of society.   Overall,
it is possible to detect a broad movement from
individualism, through utilitarianism, towards what

may best be described as socialism. Underlying this tendency was the growing belief that the individual was only partially in control of his own happiness; in other words, that he was to a large extent influenced by his environment (in the broadest sense of the term). This ideological development, it has been argued, generated a growing willingness to restrict individual freedoms in the interests of society as a whole, or at least of certain groups within it (Collini, 1976). However, historians have encountered great difficulty in linking such ideological change to specific changes in social policy. A connection is particularly hard to establish in the area of urban environmental policy, much of which developed as the result of local initiatives by individuals who were often apparently unaware of the philosophical implications of their actions (McCord, 1978). Indeed, it is often argued that attitudes and beliefs directly capable of generating social policy changed very little during the century (Thane, 1978).

With our problem still largely unresolved, we must turn to the *technological* explanation. Here it is assumed that rising technical competence in nine-teenth-century towns tended to extend the area of activity conceded to public authorities without requiring important ideological changes (Kellett, 1978). Completely new elements were added to the infrastructure, including high-pressure water supplies, gas, water-borne sewerage, railed transport, electricity, and telephones. Medical knowledge was also extended, thus generating a wider awareness of potential threats to the happiness and health of the urban community, while new techniques of environmental improvement such as disinfestation and disinfection were also developed. This tech-nical progress tended to reduce the utility of certain established practices of public intervention in the urban environment, and created demand for new ones which could not be undertaken efficiently by individual enterprise.

Some examples will help to clarify the argument. In a small eighteenth-century town with little wheeled traffic, the creation of adequate surfaces for thoroughfares may well have been secured efficiently by requiring all owners to pave in front of their own houses. In a large nineteenth-century industrial town, heavily dependent on the free move-ment of carts and carriages, such arrangements were likely to generate either massive inconvenience and expense, involving constant inspection and harrying of recalcitrant owners in the courts, or, alternatively, frequent variations in the road surfaces creating huge costs for individuals in terms of lost time, damaged vehicles, and general discomfort. If tram lines were added then private paving would become virtually impossible and in these circumstances the municipal paving of streets would increasingly appear as the more efficient solution. Such action would, of course, be an intensification of public intervention in the management of the urban environment, but the freedom of private interests to pursue their own ends within the town would thereby be enhanced and not restrained. Similarly, the development of waterborne sewage disposal, itself dependent on improvements in water supply, created an important new amenity which, in contrast to the previous cesspit system, could be efficiently established only by public enterprise. This technological factor can even help to explain some of the extensions of control over the use of private land; for instance, rapid improvements in building techniques would have required more intricate and restrictive building regulations, regardless of public-health considerations.

It is, however, important to recognize the limitations of the technological explanation. Many of the new facilities generated during the nineteenth century could normally have been provided by private enterprise. A charge could be made to the consumer and, indeed, without the incentive of private profit many would never have been developed at all. The subsequent extension of public control over urban facilities was the product of two separate factors. On the one hand, there was often a tendency towards monopoly, with the presumption of higher charges to the consumer (Falkus, 1977). On the other hand, the failure of certain owners and occupiers to take advantage of the facilities was increasingly regarded as detracting from the well-being of the remainder of the population. Although it is not entirely out of the question to relate these two factors and build them into a purely technological interpretation of public intervention, it is more helpful to view them in the context of the structural mode of explanation, to which we can now turn.

The *structural* explanation is based on the premise that public intervention tended to expand to fill the gap between the actual and the desired performance of the urban environment. The nineteenth-century trend towards greater intervention would thus be the product of a widening gap, the result either of an actual deterioration of the urban

112

environment combined with static or rising
expectations, or of an improving environment combined with an even faster rise in expectations.

A considerable amount of literature already
implicitly assumes the validity of such a structural
approach, within which two extreme interpretations
may be detected. On the one hand, an 'optimistic'
school of historians (Ashworth, 1954; Cherry, 1972)
presents the growth of intervention as the product
of growing awareness of the influence of the
physical environment on human well-being. Expectations first rose as the development of anti-
contagionist theories of disease over the first half
of the century generated a growing demand for a
clean environment (Pelling, 1978). Germ theories
of disease, increasingly dominant in the second half
of the century, added precision to this demand,
whilst at the same time reducing the unit cost of
effective public action. The value of sunlight and
fresh air was increasingly recognized, as were the
dangers of overcrowding. Towards the end of the
century the moral and psychological influence of the
environment came to be even more fully accepted,
thus helping to generate the embryonic town planning
of the early 1900s in which major extensions of
controls over the use of private land were tolerated.
As for the growth of public investment in the urban
infrastructure and its extension into new areas
such as public transport, the 'optimists' often
imply that growing demand from an increasingly
sophisticated public, the natural complement of the
improvement in real earnings and education, was
mainly responsible (Roach, 1978).

On the other hand, a 'pessimistic' school
(Gauldie, 1974; Wohl, 1977) presents a picture of
growing urban crisis in the nineteenth century, very
largely the product of the conditions of capitalist
production. The result was a deterioration of the
environment which affected a large and growing proportion of the urban population. This urban crisis
generated a more active intervention in the urban
environment by the directive elements of society,
either to protect themselves against the immediate
effects of general deterioration, or to control the
protests of those who were affected. Meanwhile,
the failure of the free market to stave off the
crisis undermined the dominant individualistic
ideology. Growing public intervention merely
accentuated the contradictions of the free-market
system and so accelerated its decline.

At first sight, the pessimistic interpretation
is the least plausible. The most spectacular

extensions of public intervention were made from the later 1860s, precisely when real earnings began to rise for all groups in the urban population and when the simplest indicator of urban well-being, the gross death rate, began to fall. Thereafter all the indicators suggest a tendency towards improvement in urban conditions, only part of which can plausibly be accredited to public intervention (McKeown and Record, 1962; McKeown, 1976).

However, the optimistic view also has its weaknesses. Those who espouse this explanation usually fail to take full account of the great disparities of income and power in nineteenth-century towns, and generally ignore the political implications of decisions to extend public intervention (Harvey, 1973). Moreover, they generally fail to understand the effects of the tendency for house-rents to rise as a proportion of household incomes during the nineteenth century (Weber, 1960).

So attractive is the structural mode of explanation that we ought to attempt to resolve the disagreements. We can do so by suggesting that the market mechanisms which produced the urban environment tended to decline in efficiency during the nineteenth century, so that a given quality of environment became increasingly expensive to secure relative to other goods and services. All town dwellers, except perhaps a tiny, very rich minority, would have suffered from this trend, but some would have borne a heavier burden than others. Unskilled and casual workers, together with the unemployed, would have been forced to purchase their environment on the least favourable terms of all. Eventually they may have given up the unequal struggle to improve their immediate environment, perhaps re-arranging their expenditure in ways which worried their social superiors. Thus although the urban environment may have improved overall in response to higher earnings, market malfunctionings could have prevented it from keeping up with the rapidly rising expectations of certain sectors of the population, which the same higher earnings helped to generate after mid-century.

Market forces affecting the built environment

We must now consider the ways in which the market may have declined in efficiency during the nineteenth century. First, we can note the marked tendency of urbanization to concentrate a growing proportion of the total population in a number of very large cities (Weber, 1899; Lampard, 1973).

This trend, mainly the product of external economies of scale which large centres of population and communications could offer to manufacturing and service enterprises, was reinforced from the 1840s by the creation of a national railway network within which nearly all existing large towns became nodal points. It is usual for land rents to be linked positively to the size of the town (Alonso, 1964); this concentration process thus produced an increase in average per capita urban land-rent payments (Singer, 1941, as corrected and complemented by Offer, 1980). In itself this phenomenon did not indicate a malfunctioning of the land market. However, the larger the city, the more likely it was that malfunctioning would occur, adding further, and in principle unnecessary, supplements to land rent.

Second, there is much circumstantial evidence for the development of blockages in the organisation of the building industry and in the management of rented housing during the nineteenth century (Chapman, 1971; Burnett, 1978). Let us look first at supply. Growing cities need new building land, and the largest require public transport to make peripheral land fully accessible. In theory, the supply of land and transport should respond to the development of effective demand, provided they are not controlled by monopolies. Although monopoly control of land is by no means out of the question in a small town, it is hard to envisage how it could be established in large cities such as London or Manchester. On the other hand, a sustained rise in land rent over a growing area would have tended to attract speculative capital, and thus prompt the pursuit of gain among long-established land owners (Olsen 1964, 1973). In these circumstances, some kind of combination among owners of building land could occur, thus producing a tendency towards monopoly and a further general rise in land rent. I am not suggesting that the land supply around British cities was ever restricted to the extent achieved by German speculators in the later nineteenth century (Eberstadt, 1909), but the *Spekulationsring* apparently built up around Berlin and other German cities, even in the absence of large aristocratic holdings, is an indication of the trend which the process of big-city concentration may be expected to promote.

The supply of land could be further reduced by a factor almost contrary to the above - the withholding of potential building land for non-economic motives. Certain owners, particularly those who controlled large blocks of land which they intended

115

to develop of a piece under long leases, could refuse to release land for functions which they considered undesirable. In Birmingham, for instance, the Calthorpe family developed a large estate close to the city centre solely for upper-middle class residents (Cannadine, 1977). Their policy almost certainly pushed up the cost of industrial and lower-class residential land elsewhere in Birmingham, while offering the relatively rich much cheaper housing than they would otherwise have obtained. Admittedly, there was a limit to the extent to which even the largest estates could ignore global market forces (Cannadine, 1980), but a degree of market distortion, unfavourable to poorer residents, would have arisen from the operation of this process.

In the supply of public transport there was a definite tendency towards monopoly, particularly after the superiority for most purposes of railed vehicles was established in mid-century. There is considerable circumstantial evidence to suggest that the supply of urban transport was thereby seriously restricted during the early tramway period from the later 1860s to the early 1890s (Dickinson, 1860; Ward, 1964). Excluded by law from involvement in land development, transport concerns (with the Metropolitan Railway as almost the sole exception) had no interest in emulating their North American counterparts by providing cheap transit as part of a general strategy to valorize peripheral land (Barker and Robbins, 1975). Furthermore, it is uncertain whether the eventual remedy, the establishment of municipal transport monopolies from the 1890s, succeeded in maximizing the supply of urban transport (McKay, 1976). Whereas in Germany muncipalities were allowed to engage in suburban land development (Hartog, 1962), British urban authorities were almost entirely excluded from this activity, and so faced much the same disincentives as the private transport companies in the earlier period.

The failure of public transport to expand and thus reduce urban land rents in the way it was theoretically capable, would have produced a repercussive effect in that the land needed for transport itself would have become more expensive to acquire. The use of compulsory purchase procedures, almost always necessary for transport lines, was particularly costly in urban areas because compensation for disturbance was usually higher than in the countryside. Such difficulties would have further discouraged the provision of exclusive routes for urban public transport

(Kellett, 1969). Instead, a more intensive use of existing thoroughfares, already in public ownership, became necessary, seen mainly in the form of street railways (tramways). Fundamentally unsuitable for railed transport, streets would have declined in efficiency as corridors of communication because of conflicts between different types of user. Such problems, and particularly clashes between pedestrians and electric railed vehicles using road surfaces, were accentuated as cities grew in size and complexity. There was no radical remedy for these difficulties; elevated railways built on public thoroughfares were unacceptable on environmental grounds, while tube railways, drilled in central London from the 1890s, avoided high outlay on land only by incurring very high construction costs.

It is one of the ironies of nineteenth-century urban history that this rise in land costs was not counteracted by a fall in the construction and management costs of buildings (Maiwald, 1954). Unfortunately, these generally failed to benefit from the external economies of scale generated by increasing urban concentration. As economic development proceeded, residential building became less attractive to capital, while at the same time the construction industry largely failed, particularly in house-building, to develop new techniques and organization capable of increasing its productivity. At the bottom of the rental market, the supply of housing was further reduced by a growing unwillingness among owners to manage accommodation let to the very poor. As a result, demand for the cheapest accommodation was increasingly satisfied, especially in London and no doubt in other large cities as well, by a class of low landlords, factors or 'house-farmers' who operated in more or less tightly organized rings, with their position sometimes reinforced by dominance of the local authority (Jones, 1961). These low landlords became particularly involved in letting to poor tenants the stock of larger, older houses, which had been forsaken by their original superior class of occupant. In this housing market supply was thus distorted, not so much in terms of quantity, which was considerable, but in terms of quality. The division of these larger houses into flats and rooms appropriate to the means of poorer occupants nearly always involved a very sharp reduction in the quality of both the private environment and immediate public environment.

In looking only at the supply side, we have already seen an increase in the production cost of the

117

urban environment suffered by all except those very rich who were able to live on cheap land outside the town, and who could provide most of their own services, including transport. Counterbalancing the cheap environment of the rich was the inordinately expensive environment of the very poor. At the same time, the poor also suffered particularly seriously from distortions of demand which almost certainly grew more acute in the later nineteenth century. The fundamental problem in this regard was the inefficient functioning of the labour market (Jones, 1971). The social attractions of urban life, combined with relative ease of migration, may have tended to produce an over-supply of labour in most large cities. In theory, the concomitant relative under-supply of labour in certain parts of Britain and abroad should have produced a redistribution of the population, but it seems reasonable to suppose that residents of large cities were insufficiently aware of opportunities elsewhere to make such moves. This immobility particularly affected the growing proportions of people who had been born and brought up in the largest towns and who were thus perhaps more insular and therefore less likely to settle easily elsewhere.

Over-supply of labour was further accentuated by the desperate efforts of family-members to supplement low earnings of the main breadwinner. Multiple unskilled and casual employment within a family would have discouraged efforts to seek cheaper housing in outer districts of the city which were distant from most employment opportunities. The larger the city, the more this effect would be reinforced by the growing social isolation of poor neighbourhoods as segregation of functions increased during the century. Not merely ignorant of opportunities elsewhere in the town, slumdwellers may well have been positively attached to the rude, vibrant quality of life in districts where lack of residential mobility and high densities generated a personal intimacy which was reinforced by complex kinship ties (Dyos, 1967; Roberts, 1971; Hoggart, 1958; Wilmott and Young 1962). Industries capable of employing the lowest grade of labour, such as the sweated trades, were also attracted to the inner districts (Hall, 1962), further discouraging workers from undertaking a fundamental reconsideration of their residential location. The scissors effect of low wages and high rents was in itself serious, but it also led to a final closing of the vicious circle through voluntary restrictions in effective demand for housing.

Discussion

What follows is more speculative, but nonetheless
plausible. Offered such unfavourable terms, the
poor town-dweller might quite reasonably have given
up any efforts to secure a more pleasant urban
environment, choosing instead to devote his meagre
resources to the purchase of goods or services
available on more favourable terms. With food,
clothing and entertainment growing cheaper during
the later nineteenth century, he could have compensated for his poor environment in many different
ways. The strongest temptation, however, may have
been to limit his perception of the unpleasant
environment by consuming alcohol. A degree of ineb-
riation would not have debarred the labourer or the
outworker from employment; indeed, labourers such
as coal-heavers were often encouraged to drink in
order to combat dehydration. Beer and spirits,
which could be consumed at home, would have diverted
attention from a poor living environment more
easily than outside distractions such as the music
hall. Furthermore, alcohol would have been
particularly cheap in the inner districts, owing to
a concentration of retail outlets. We may even
speculate the existence of an urban drink-price
curve which, in contrast to rents, would have risen
towards the outer districts, thus discouraging the
devotion of personal resources to drinking in the
suburbs.

This whole model, if acceptable, makes
deterioration in conditions of production of the
urban environment the fundamental social problem of
the later nineteenth century. This may seem pre-
posterous, but the model is not merely the product
of detached theorizing. It leaps from the writings
of a whole wave of social analysts and reformers
between the 1870s and 1914, including Octavia Hill
and Patrick Geddes. Indeed, the thrust of many
reform proposals of the period was towards adjustment of the urban structure. Some favoured redistribution of population towards the outskirts of
cities, stimulated by cheaper transport or cheaper
land. Others wanted to remove surplus population
from the cities altogether, through the provision of
industrial villages, labour colonies, garden cities,
and assisted or even compulsory emigration to the
Empire (Harris, 1972). A further lobby favoured
direct public action to bypass blockages in the construction industry by municipal house-building, and
a small minority was even prepared to consider

119

subsidization of rents.   Others, for the most part
distrustful of subsidization, wanted to encourage a
shift in personal resources towards housing by local
regulations designed to require a more attractive
product from private developers and builders.
Finally, public ownership of peripheral building land
was increasingly canvassed, as part of a general
land-reform movement which grew in strength towards
the end of the nineteenth century.

The ultimate product of these diagnoses was the
idea of town planning, which emerged in the early
1900s.   It incorporated, in a single strategy, the
entire range of specific public interventions, both
positive and negative, which could be used in the
production of the urban environment.   The writings
of early planning ideologues, such as Nettlefold
(1908) and Geddes (1915), revealed a firm conviction
that the social malaise of the late-nineteenth and
early-twentieth centuries was the result of a breakdown in the production of the urban environment.
They were confident that, once the various blocks
and distortions had been removed, the existing social
and economic system would once again generate
progress towards a happier human condition.

History has not dealt kindly with this
diagnosis, and social reform has for the most part
followed a quite different course to that advocated
by Nettlefold.   Certainly, the diagnosis had its
weaknesses.   Adjustment of the urban environment
alone could not have eradicated poverty.   A more
healthy, uplifting environment such as Port Sunlight
or Bournville would not have produced a more culti-
vated or cooperative workforce.   However, as the
basis for a reform policy, the analytical weaknesses
of the diagnosis were largely compensated for by its
political attractiveness.   The same programme could,
or so it seemed at the time, eradicate working-
class discontent, create a more efficient workforce,
and ease the task of the middle classes in securing
a pleasant environment for themselves.   It would be
paid for, not by redistributive taxation, but by the
removal of various friction costs (Unwin, 1913), or
by the partial expropriation of a small and unpopular
class of landowners.   No ideological changes would
be required.

Although this diagnosis and the solutions
reached their mature form only in the early 1900s,
the trend towards them was apparent over at least
the preceding thirty years.   I would seriously
suggest that the development of intervention beyond
the largely sanitary stage which lasted from the
1840s to the 1870s was in large measure a response

to the structural malfunctioning in the urban
environment which I have described.

Conclusion

What then are the implications of this for future
research? First, we need to look more closely at
the grounds on which extensions of public intervention in the urban environment were justified in
individual towns.   This will entail looking closely
at the multitude of local act proceedings, and at
court cases in which local powers were challenged
(Barber, 1980).   Second, we need to look systematically at the ways in which local-government areas
were defined, much as our German colleagues are
already doing (Matzerath, 1978).   How far, for
instance, were rich refugees from the towns able to
create their own environment, and do without a full
range of municipal services? To what extent did
they wish to do so?   Third, we must devote more
attention to evaluating the costs of suburban middle-
class housing, in relation to the quality of
environment achieved.   Were the middle classes, for
instance, really obtaining less and less for their
money as time went by?   Fourth, we need to follow
the example of Gareth Stedman-Jones by studying in
detail the local economy and society of slum areas
in conjunction with the environmental conditions
within them.   Fifth, we must consider much more
seriously the whole question of urban land-rent and
the land-reform movement of the later nineteenth
century (Offer, 1980).   Sixth, we could learn much
by studying those episodes in which the trend
towards public intervention was apparently reversed,
such as Birmingham's withdrawal from municipal
house-building in the early 1900s (Sutcliffe, 1974).
Finally, and most important of all, we must study
the development of environmental intervention in any
chosen town in association with the evolution of
that town's economic, social and physical structures.
We shall never properly understand the progress of
urban policy until we relate it fully to the
processes of urban change.

References

    Alonso, W. (1964) *Location and land use: toward a general
theory of land rent*, Harvard University Press, Cambridge, Mass.
    Ashworth, W. (1954) *The genesis of modern British town
planning: a study in economic and social history of the*

121

nineteenth and twentieth centuries, Routledge and Kegan Paul, London

Barber, B.J. (1980) 'Aspects of municipal government, 1835-1914', in D. Fraser (ed.) *A History of Modern Leeds,* Manchester University Press, Manchester, pp.301-26

Barker, T.C. and Robbins, M. (1975 revised edition) *A history of London Transport: passenger travel and the development of the Metropolis. Vol. I: the nineteenth century,* Allen and Unwin, London

Burnett, J. (1978) *A social history of housing, 1815-1970,* David and Charles, Newton Abbot

Cannadine, D. (1977) 'Victorian cities: how different?' *Social History,* 4, 457-82

Cannadine, D. (1980) *Lords and landlords: the aristocracy and the towns, 1774-1967,* Leicester University Press, Leicester

Chapman, S.D. (1971) *The history of working-class housing: a symposium,* David and Charles, Newton and Abbot

Cherry, G. (1972) *Urban change and planning. A history of urban development in Britain since 1750,* Foulis, London

Chinnery, G.A. (ed.) (1974) *Records of the Borough of Leicester. Vol. VII. judicial and allied records 1689-1835,* Leicester University Press, Leicester

Collini, S. (1976) 'Hobhouse, Bosanquet and the State: philosophical idealism and political argument in England, 1880-1918' *Past and Present,* 72, 86-111

Cromwell, V. (1966) 'Interpretations of nineteenth-century administration: an analysis', *Victorian Studies,* 9, 245-58

Dickinson, G.C. (1960) 'The development of suburban road passenger transport in Leeds, 1840-1895', *Journal of Transport History,* 4, 214-23

Dyos, H.J. (1967) 'The slums of Victorian London', *Victorian Studies,* 11 (supplement), 5-40

Eberstadt, R. (1909) *Handbuch des Wohnungswesens und der Wohnungsfrage,* Gustav Fischer, Jena

Falkus, M. (1977) 'The development of municipal trading in the nineteenth century', *Business History,* 19, 134-61

Gauldie, E.(1974) *Cruel habitations: a history of working-class housing 1780-1918,* Allen and Unwin, London

Geddes, P. (1915) *Cities in evolution: an introduction to the town planning movement and to the study of civics,* Williams and Norgate, London

Hall, (1962) *The Industries of London since 1861,* Hutchinson, London

Harris, J. (1972) *Unemployment and politics: a study in English social policy, 1886-1914,* Clarendon Press, Oxford

Hartog, R. (1962) *Stadterweiterungen im 19. Jahrhundert,* Kohlammer, Stuttgart

Harvey, D. (1973) *Social Justice and the City,* Edward Arnold, London

Hoggart, R. (1958) *The uses of literacy: aspects of working-class life with special reference to publications and entertainments,* Penguin, Harmondsworth

Jones, G.S. (1971) *Outcast London: a study in the relationship between classes in Victorian society*, Clarendon Press, Oxford

Kellett, J.R. (1969) *The Impact of Railways on Victorian Cities*, Routledge and Kegan Paul, London

Kellett, J.R. (1978) 'Municipal socialism, enterprise and trading in the Victorian city', *Urban History Yearbook*, 36-45

Lampard, E.E. (1973) 'The urbanizing world', in H.J. Dyos and M. Wolff (eds.), *The Victorian City: images and realities*, Routledge, London, pp.3-57

McCord, N. (1978) 'Ratepayers and social policy', in P. Thane (ed.) *The origins of British social policy*, Croom Helm, London, pp.21-35

McKay, J.P. (1976) *Tramways and trolleys: the rise of urban mass transport in Europe*, Princeton University Press, Princeton

McKeown, T. (1976) *The Modern Rise of Population*, Edward Arnold, London

McKeown, T. and Record, R.G. (1962) 'Reasons for the decline in mortality in England and Wales during the nineteenth century', *Population Studies*, 16, 94-122

Maiwald, K. (1954) 'An index of building costs in the United Kingdom, 1845-1938', *Economic History Review*, 2nd series 7, 187-203

Matzerath, H. (1978) 'Stadtewachstum und Eingemeindungen im 19. Jahrhundert', in J. Reulecke (ed.), *Die Deutsche Stadt im Industriezetalter*, Peter Hammer Verlag, Wuppertal, pp.67-89

Nettlefold, J.S. (1908) *Practical Housing*, Garden City Press, Letchworth

Offer, A. (1980) 'Ricardo's paradox and the movement of rents in England, c.1870-1910', *Economic History Review*, 2nd series, 33, 236-52

Olsen, D.J. (1964) *Town planning in London: the eighteenth and nineteenth centuries*, Yale University Press, New Haven

Olsen, D.J. (1973) 'House upon house', in H.J. Dyos and M. Wolff (eds.) *The Victorian City: Images and Realities*, Routledge, London, pp.333-57

Pelling, M. (1978) *Cholera fever and English medicine 1825-1965*, Oxford University Press, Oxford

Roach, J. (1978) *Social reform in England 1780-1880*, Batsford, London

Roberts, R. (1971) *The classic slum: Salford life in the first quarter of the century*, Manchester University Press, Manchester

Singer, H.W. (1941) 'An index of urban land rents and house rents in England and Wales, 1845-1913', *Econometrica*, 9, 221-30

Sutcliffe, A. (1974) 'A century of flats in Birmingham, 1875-1973', in A. Sutcliffe (ed.), *Multi-Storey living: the British working-class experience*, Croom Helm, London, pp.181-206

Thane, P. (1978) 'Introduction', in P. Thane (ed.) *The*

origins of British social policy, Croom Helm, London, pp.11-20

Unwin, R. (1912) *Nothing gained by overcrowding: how the garden city type of development may benefit both owner and occupier*, P.S. King and Son, London

Ward, D. (1964) 'A comparative historical geography of streetcar suburbs in Boston, Massachusetts and Leeds, England: 1850-1920', *Annals of the Association of American Geographers*, 54, 477-89

Webb, S. and Webb, B. (1922) *English local government: statutory authorities for special purposes*, Longmans, London

Webb, S. and Webb, B. (1913) *English local government: the story of the King's highway*, Longmans, London

Weber, A.F. (1899) *The growth of cities in the nineteenth century: a study in statistics*, Macmillan, New York

Weber, B. (1960) 'A new index of house rents for Great Britain, 1874-1913', *Scottish Journal of Political Economy*, 7, 232-7

Wohl, A.S. (1977) *The eternal slum: housing and social policy in Victorian London*, Edward Arnold, London

Young, M.D., and Wilmott, P. (1962) *Family and kinship in east London*, Penguin, Harmondsworth

PART THREE

RETAILING AND THE NINETEENTH-CENTURY
URBAN ECONOMY

The economic structure of nineteenth-century cities
remains something of an enigma.   Most of the major
urban manufacturing industries have had their
biographers, but these histories have been primarily
concerned with the development of the individual
firm or industry.   Only incidentally have they
explored the links between industrialization and
urbanization, and the effects of industrial and
economic change on the internal structure of the
town.   Furthermore, methods of recruitment and the
development of the nineteenth-century urban labour
market have rarely been explicitly examined.   There
is thus much scope for the further exploration of
the specifically urban aspects of nineteenth-century
economic development and of the effects of economic
change on nineteenth-century urban structure.
Unlike manufacturing industry, nineteenth-century
urban service occupations have been neglected on
most counts.   There are few studies of firms involved in providing services and the effects of the
growing service and commercial sectors on urban
structure have been little considered until
recently.
     The three essays in this section attempt to
explore only one aspect of this vast topic: the
development of the retailing sector in the nineteenth-century urban economy.   Although it may be
criticized for ignoring manufacturing industry and
the rest of the service sector, the strategy
adopted has been to focus on one relatively-
neglected area of the urban economy.   If it is
possible to demonstrate that this sector has un-
tapped potentialities for research, particularly
into the links between economic changes and the
evolving internal structure of cities, then in the
future the extension of similar work to other
sectors of the urban economy may be encouraged.

David Green's chapter focuses on the army of
itinerant traders that filled the streets of nineteenth-century London, viewing these people as being
employed in one tier of the nineteenth-century urban
labour market. The casual street trader was
typified by economic, legal and spatial marginality,
but this sector of the labour market was a vital
part of both urban service and employment provision
in Victorian London. The problems of studying
casual street traders are immense, not least
because the paucity of data makes it difficult even
to make a realistic estimate of the total size of
the urban service sector. Conventional census
sources severely underestimate the size of the
casual labour force, but through the use of Poor Law
Records and a range of other archival material
Green begins to reconstruct something of the
importance of street trading both as a source of
income and employment for the urban poor, and as a
system of supply within the urban service sector.

The other two chapters deal mainly with the
formal side of urban retailing, particularly the
rate of shop development, the locational decisions
of shop-keepers, and the trade areas of different
types of retail establishments. There are many
points of contact in the contributions of the
historian (Roger Scola) and the geographer (Gareth
Shaw). Both stress the relative neglect of the
retailing sector in conventional studies of urban
development, and both emphasize the problems
inherent in the sources available, which so far have
been an inhibiting factor for such studies. There
are also important differences between these two
essays. The authors arrive at rather conflicting
estimates of the rates of shop growth and the
extent to which retailing kept pace with population
growth, no doubt reflecting the problems of source
interpretation. While Roger Scola begins to apply
geographical techniques of locational analysis to
his study of Manchester, Gareth Shaw takes this
theme much further, using a number of indices of
retail development on a range of nineteenth-century
towns.

One obvious question arises: to what extent
are the techniques and methods developed for
studies of modern urban geography applicable to
the Victorian town? Not only were economic
circumstances and consumer behaviour likely to be
different, but the severe constraints imposed by
available sources of information may also make many
techniques totally inapplicable. Although there
are many problems it is clear that there are

considerable opportunities for more detailed
analysis of urban retailing (and other aspects of
the urban economy).

Chapter 5

STREET TRADING IN LONDON: A CASE STUDY OF CASUAL LABOUR 1830-60

David R.Green

Introduction

At the hub of the expanding urban network of nineteenth-century Britain lay London, attracting both
the nation's capital and population. At the start
of the century it was a city of under a million inhabitants, but by 1851 it had increased to almost
two and a half million (Price-Williams 1885, p.380).
Stimulated by railway links to virtually every town
of any importance (Dyos and Aldcroft, 1974, p.145),
London's growth rate between 1841-51 soared to 21.24
per cent, the highest in the century.   Indeed, the
Great Exhibition of 1851 almost seemed to suggest
that the world itself had been attracted to the
capital.
     Widespread and rapid urban growth generated a
whole new set of social conditions and problems.
Above all, perhaps, it called forth the urban crowd.
In London the city mob had been particularly active
during the eighteenth century, reaching its climax
in the Gordon riots of 1780 (Rudé, 1974, pp.268-92).
During the nineteenth century the activities of the
crowd most prone to be historically significant took
the form of strikes, mass meetings and political
demonstrations (Rudé, 1964, p.5), but this was the
crowd at its most sensational and ephemeral.   The
new urban experience was a far more commonplace one
of crowded streets and congested thoroughfares.
Reminiscing about his first visit to London in the
late 1820s, James Burn wrote: 'I was fairly lost in
a wilderness of human beings; I was a mere atom in
a huge mountain of humanity' (Burn, 1855, p.95).
The young Friedrich Engels, in London for the first
time in 1843, also found 'something repulsive'
about the turmoil of the streets, the 'brutal indifference' and 'unfeeling isolation' of the urban
crowd whose 'only agreement is the tacit one; that

each keep to his own side of the pavement, so as not to delay the opposing streams of the crowd...' (Engels, 1973, p.60).

If the crowd provided a new urban experience, then the street trader was one of its primary characters. Time and again visitors to London, as well as residents, commented on the large number of itinerant traders of all sorts that wandered about the streets (Engels, 1973, p.122; Kirwan 1963; Babbage, 1864). Far from being quaint exotica, the street traders - costermongers, flower girls, street musicians, even prostitutes - were as much rooted in the casual labour market as the more commonly quoted examples of dock labourers and building workers. With this in mind the intention of this chapter is twofold: first, to demonstrate how an increase in the numbers of street folk stemmed from the structural conditions inherent in the casual labour market; and, second, to examine how this increase affected the extent of street trading and led to a series of middle-class attempts at control and re-pression.

Street traders and the urban labour market

London lacked any real sense of metropolitan unity in the first half of the nineteenth century. As one contemporary author put it, inhabitants of neigh-bourhoods separated by the width of a street 'for all they know of one another are at as great a distance as the natives of China and Peru' (Murray, 1843, p.20). As well as reflecting the growing separation of rich and poor, and the complex internal residential differentiation of the less affluent majority (Ward, 1976, p.330), this state-ment highlights the intense disaggregation of the London labour market. As Stedman-Jones has pointed out, small-scale production carried on in the work-shop or in the home, coupled with the relative absence of the factory, meant that the social character of London's labouring population was peculiarly individual (Stedman-Jones, 1971, p.31). This pattern was the outcome of attempts to compete with cheaper provincial factory production, growing in importance from the 1860s but nevertheless significant by mid-century. By reducing workshop production to a minimum and expanding work at home, employers sought to diminish the two main London overheads of rent and wages. The dilution of trade skills allowed employers to reduce wage rates. At the same time the transfer of production to the home

130

permitted the domestic labour of wives and children
to be called upon.   The outcome was a proliferation
of small units of production in which labour power
was substituted for that of machines.   Thus in 1851,
86 per cent of London employers hired less than ten
men each and the majority employed less than five
workers.[1] This situation, graphically described by
Mayhew in his letters to the *Morning Chronicle* in
1849-50 (Thompson and Yeo, 1973), existed chiefly in
the traditional London consumer trades, notably
clothing and woodworking (Hall, 1962).   These could
be found in the districts immediately surrounding
the City and it was in this inner industrial zone
that the problem of casual labour assumed its most
significant form.

For the skilled artisan, local employment
districts (representing a worker's home area) were
part of a wider pattern of regional labour markets.
Hobsbawm (1964) suggests that three such markets
existed in London, each preserving its own autonomy
and wage levels whilst reflecting real divisions
between working-class areas of the city.   For casual
labour, however, the necessity of being on hand
should demand arise effectively tied workers to
central locations.   In addition, the opportunities
for unskilled female labour became concentrated in
the districts of the inner industrial area.   This
added a further constraint to the low geographical
mobility of casual labour already engendered by
poverty, ignorance and the importance of local ties
in the community (Stedman-Jones, 1971, pp.81-8).

In these districts there existed a large pool
of unskilled labour, tied to the locality but rarely
employed with any degree of regularity.   Indeed,
with low fixed capital investment and this constant
labour surplus, employers were able to adjust their
workforce according to the state of demand.   Consequently casual labour was also characterized by
only short periods of employment.   Whilst irregular
demand was met by brief, often occasional employment,
which ranged from hours or days up to a 'job',
seasonal fluctuations were of more general significance.   For the small master stockpiling of goods
was impossible, and therefore seasonal slack periods
in the casual labour market resulted in widespread
unemployment.   This tended to occur during the
winter months.   Seasonality, however, merely reinforced the tendency for casual labour to be a form
of chronic underemployment (Beveridge, 1909, pp.102-
9).   For the labourer employment was both unpredictable and intermittent and consequently the
average income was often insufficient for subsis-

tence. To supplement the family budget the
domestic labour of wives and children were often
resorted to (Alexander, 1976, pp.97-110; Lees, 1976,
pp.35-7). Not only did this further restrict the
casual labourer to a particular district but it had
the tendency to reduce wage rates and further over-
stock the labour market (Mayhew, 1861, vol. 2, p.313).
    One of the most obvious manifestations of this
chronically under-employed labour force was the
crowd of persons that thronged the street. It
greeted the urban bourgeoisie first thing in the
morning, disturbed it at all hours of the day and
left it last thing at night (Smith 1857, pp.356-99).
One of the most visible and audible members of this
throng were the London street traders or those who
'picked up a crust' in the public thoroughfares.
But this had not always been the case. In the
early decades of the century, many hawkers from the
countryside around London trudged to the city to
sell their wares. 'Simplers' supplied the markets
with herbs. William Friday, for instance,
regularly walked to London from Croydon, first
selling the mushrooms he had picked at the start of
his journey and then becoming in turn 'snail picker',
'leech bather', and 'viper catcher' (Smith 1817,
p.44). In the 1820s women could be seen selling
vegetables and fruit freshly gathered from market
gardens in the suburbs (Anon, 1823, p.10). The
rabbit and poultry seller who wandered down
Piccadilly in the 1850s had purchased his goods at
Leadenhall market (Leighton, 1851, p.16), whereas
his predecessor in the 1820s had wandered in from
the country suburbs to hawk the game which he had
caught himself (Sam Syntax, 1820, pp.11, 15). Most
telling of all, however, was 'Honest Jack' the
basket maker. In the 1820s he could be found around
the Thames between Fulham and Staines 'near meadows
damp and low, where his bending osiers grow' (Anon,
1823 p.44). But by 1839 'Honest Jack' had been displaced from his cottage by an expanding urban
middle-class in search of rural retreats in which
to locate their fashionable villas. 'Jack', we are
told, could now be found in a damp cellar off the
Haymarket (Smith 1839, p.56).
    The expansion of the urban fringe profoundly
altered the work involved in urban hawking and helps
to explain why such characters as 'Honest Jack' and
William Friday were replaced by the brooding and
melancholy woodcuts of Mayhew's street sellers.
From being food producers in their own right, urban
costermongers were metamorphosed into a link in the
chain of consumption. They were obliged to rely on

the central markets for supplies of vegetables, fruit and fish. As such they lost direct control over the supply of goods, which in turn came to be dictated by market forces. In effect costermongers came last of a number of potential buyers when the goods came to be sold at the central markets. In Covent Garden, the most important market for coster-mongers in the west of London, salesmen or 'higglers' purchased whole cartloads of produce direct from country growers and by acquiring a monopoly over the sale of goods, they were able to force up prices[2]. Since costermongers relied on purchasing large quantities cheaply they bargained from a particularly weak position. Indeed, at times they were obliged to trade with higglers who sold short measures[3]. In Spitalfields, the major East End market, coster-mongers had to wait until 9 a.m., after the green-grocers had already made their purchases, to buy what was left[4]. Similarly, in Billingsgate sales-men refused to accept a bid below an agreed fixed price[5]. What hope could there possibly be for the fish hawker when even Baroness Burdett-Coutts' Columbia Market foundered when opposed by the Billingsgate salesmen (Stern, 1966, pp.357-9).

But how were these urban street traders related to the casual labour market? This was a question to which Mayhew repeatedly referred throughout the four volumes of *London Labour and the London Poor*. In an obvious sense the street traders themselves provide us with an answer. According to Mayhew, three groups of people became street traders: those who were bred to the streets; those who took to them for love of the wandering life; and those who were driven there (Mayhew, 1861, vol. 1 pp.320-23). It was this latter group which provided the recruits for an 'extraordinary increase' in the late 1840s of street traders in London (Mayhew, 1861, vol. 2 p.5). 'Beaten-out' mechanics and unemployed artisans, rather than go to the workhouse, resorted to earning a living in the streets. Indicative of this group was John Jennery, a middle-aged box maker. When brought before the magistrates for hawking without a licence, he stated:

> I am starving, as well as my family of five children. I took this, the only box I had, out to sell, and was unsuccessful....Starvation stares us in the face. Everything I possess in the world is not worth two pounds.[6]

Similarly, in 1832 David Hammond, a mattress maker who had been unable to regain his former employment

after illness, took to hawking some boxes made by
his brother[7].

Mayhew was not alone in his assessment of the
situation.   Some years before Mayhew wrote his
letters to the *Morning Chronicle,* Engels had noted
that it was the 'so called "surplus population"',
the industrial reserve army which was only briefly
kept employed during boom periods, that took to
'huckstering' (Engels, 1973, p.112).   Likewise for
Thomas Archer it was one of the resorts for the
pauper population:

> (P)eople in whose families pauperism, that is
> to say, pauperism with short alternations of
> semi-starvation out of the workhouse, has
> become the ordinary condition.   Their lives
> out of the workhouse are supported by casual
> labour.   The men often work at the docks; or
> at chairmending, and deal in such small wares
> as can be sold from a huckster's barrow in poor
> neighbourhoods, or follow any chance calling
> for which they are strong enough.   The women
> and children make lucifer-match boxes; plait
> straw matting, make baskets.   Many of them
> either work or cadge indifferently, live anyhow
> or nohow, and sometimes, but not very often
> steal (Archer, 1865, p.53).

It was not just cyclical and structural unemployment
that provided the street trades with a growing
number of recruits.   Seasonal slack periods left
many labourers out of work.   During the winter
months bricklayers' labourers took to hawking hot
potatoes and chestnuts (Mayhew, 1861, vol. 1, p.175),
whilst Irish dockers and hodmen joined their wives
and children in selling oranges and nuts (Mayhew,
1861, vol. 1, p.105).   Towards the end of the year,
after the hop-picking season, many tramps and
vagrants flocked to London, also swelling the
numbers of street traders.   This general influx
into the trade occurred at a time when there were
fewest customers.   Consequently even the elite
amongst costermongers had difficulty in tiding over
the winter months, as evidenced by the creation of
a winter loan fund by the Friendly Association of
London Costermongers in 1850[8].
    We are fortunate in being able to examine the
general relationship between street trading and unemployment in some greater detail for St. Giles,
using Settlement and Examination books compiled by
the poor law authorities.   These books recorded
details of every applicant who came before the

Assistant Overseers of the Poor and an examination of the pauper's circumstances was a necessary concomitant to receiving any relief. In St. Giles the books were first kept on a daily basis, but as the poor law administration was rationalized, the days on which applications were dealt with were reduced to three. These records, therefore, not only provide a very comprehensive set of personal details but also mirror the day-to-day fluidity of lower-class life that is missed by the decennial census. St. Giles, too, as well as being one of the most notorious rookeries in mid-nineteenth century London, was a district noted for its street traders and Irish population[9]. Not only was Covent Garden close by, but a host of other small street markets were within easy reach. Consequently the area itself is particularly suitable for an examination of street trading and the casual labour market.

Tables 5.1 and 5.2 were compiled from a sample drawn from the Settlement and Examination books between 1830 and 1860. The occupational breakdown they reveal tells a familiar story of the decline of traditional London consumer trades and a highly restricted market for female unskilled labour. The largest proportion of men came from general unskilled work. This group comprised casual labourers such as porters, cabmen and bricklayers' labourers. If the traditional London trades of building, metalwork, clothing and shoemaking, and woodwork are combined they provide almost a third of all male applicants throughout the period. Again, this reflects the general situation that existed in London during the period under consideration. Coach building, located in Long Acre, also began to decline in these years. Street trading, meanwhile, provided a small but significant proportion of male applications between 1830-44. However, in the second half of the period it increased in importance both in relative and absolute terms. This finding lends weight to the observations made by Mayhew and some of his contemporaries. As well as pointing to the growing importance of the street trades as the last shift before the workhouse, it hints at a general increase in the numbers actually involved.

Street trading was of even greater significance for women. Over a quarter of female applicants for relief between 1830-44 were recorded as being street traders, and although this figure declined in the following years it nevertheless remained a significant proportion. The very limited opportunities for unskilled female labour are highlighted by the fact that five occupational groups, including

Table 5.1: St. Giles 1830-60: Occupational groups of male applicants for poor relief*

	1830 – 44 (%)	1845 – 60 (%)
Clothing and Shoemaking	16.3	14.5
Metalwork	7.3	6.1
Building	6.9	6.5
Woodwork and furniture	6.4	7.3
Coach building	6.4	1.9
Miscellaneous manufacture	4.7	4.2
Domestic service	7.7	3.8
Services and retailing	6.0	5.7
Street trading	8.2	11.8
Clerical and white collar	2.6	2.7
General unskilled	27.5	35.5
	100%	100%
	(N=233)	(N=262)

* Figures calculated from a 1 in 20 sample of Settlement and Examination books 1830-60 for St. Giles and St. George's Bloomsbury poor law district.

Table 5.2: St. Giles 1830-60: Occupational groups of female applicants for poor relief*

	1830 – 44 (%)	1845 – 60 (%)
Street trading	26.6	18.4
Domestic service	21.3	26.0
Charring and washing	19.6	20.0
Needlework	17.3	22.6
Prostitution	6.1	6.8
Shoebinding	2.0	3.0
Others	7.1	3.2
	100%	100%
	(N=342)	(N=234)

* Figures calculated as in Table 5.1.

prostitution, provided over 90 per cent of all
applications during the period.   Indeed, many of
the wives of Irish labourers had even less choice
since, as Mayhew pointed out,'A needle is as useless
in their fingers as a pen.' (Mayhew, 1861, vol. 1,
p.105).   The 'poor basket woman' was a common
feature of mid-Victorian London and Mayhew estimated
that between 25,000 to 30,000 wives, widows and
single women were engaged in the street trades
(Mayhew, 1961, vol. 1, p.463).   If any tentative
conclusions can be drawn regarding the numbers
actually involved, it is that at least as many
women as men earned a living from the streets.
        Precise estimates of the extent of street
trading are impossible to make.   The 1841 census
enumerated 2045 hawkers, hucksters and pedlars, and
the figure for 1851 was 3723.   Undoubtedly these
figures are serious under-estimates and Mayhew
treated them with extreme scepticism.   His estimate
for the late 1840s was about 30,000 adults earning
a living by street trading.   However, Mayhew's
figures are notoriously ambiguous (Thompson, 1967,
p.58) and he later stated that there were at least
25,000 women street traders.   In total he considered that a figure upwards of 45,000 would not
have been far from the truth (Mayhew, 1861, vol. 2,
p.1).   On the other hand, James Burn, himself a
pedlar, estimated that about 100,000 people earned
their living in the London streets (Burn, 1858,
p.35).   These varying estimates partly reflect
different perceptions of what constituted a street
trader, although it was obvious to contemporary
observers that the numbers were increasing.   For
example, the influx of Irish immediately following
the Great Famine was thought to have doubled the
already considerable numbers of Irish street
traders.   But it is also likely that the
variation in the estimates had a greater significance since it confirms the view shared by Engels,
Mayhew and Archer, that street trading was often an
occupation resorted to by the 'surplus population'.
For some, it was a full-time occupation and it is
this proportion that is probably recorded in the
documents, but it was also an intermittent pursuit
taken up by casual labourers and their families in
order to supplement inadequate or irregular incomes.
The variation, then, reflects a situation in which
a permanent core of street traders was joined at
times by an equally-large temporary influx.   The
size of this influx depended upon the structural
conditions which existed in the casual labour
market and the extent of cyclical and seasonal

unemployment.

Middle-class control of street trading

The increasing numbers of individuals that took to
the streets for a living was a direct result of the
structural condition of the London labour market.
But it was not the only type of increase. It was
painfully obvious to the urban middle-class that
street traders not only appeared to be ubiquitous
but were heard to be so. London hawkers, 'cried'
their goods, so that their presence on the London
streets was augmented by a rare audibility. A
'Constant Reader' of *The Times* painfully remarked
on New Year's Day, 1845, that the sound of barrel
organs and street music could be heard between the
hours of 7 a.m. and 11 p.m.[10] In 1856 *The Times*
remarked how cries of 'Dust ho', 'Water-creases' and
'Mackerel' could be heard in quiet suburban streets
starting at 8 a.m. and continuing throughout the
day[11]. It was precisely at these early hours, of
course, that shops were still shut. At night, when
many of them remained open, complaints about street
noise were usually directed not against costermongers but against Italian organ grinders and other
musicians.
Complaints against the 'crying evil' and the
'organ nuisance' continued and in November 1857
The Times remarked that:

> Hawkers succeed each other in such regular and
> frequent fashion that no street is ever
> without one shouting at the top of a practised
> voice stale lettuces, or fish or lobsters.[12]

The inhabitants of fashionable Belgravia petitioned
Parliament to treat street musicians as beggars[13]
and in 1858 ratepayers in Paddington[14] and
Westminster[15] complained of street cries and music
in their quiet streets. Charles Babbage conducted
a campaign against the 'organ nuisance' and
recorded that from 3rd July 1860 to 1st May 1861 he
had been disturbed no fewer than 165 times, mostly
by Italian organ grinders (Bass, 1864, pp.20-22).
Throughout the decade complaints against street
noise mounted and reached a crescendo in 1864 with
the passing of Bass's Act for the 'Better Regulation of Street Music Within the Metropolis'.[16] But
it seems unlikely that quite suddenly the Victorian
ear had become perceptibly more sensitive. Rather
it confirms the impressions of Mayhew and others

138

of an increase in the numbers of street traders
from the late 1840s and 1850s.

Despite these observations, increasing street
noise was not solely the result of a corresponding
increase in street traders. It also reflected two
further factors: a lengthening of the hours worked
by costermongers and the spread of street trading
into suburban districts. Costermongers interviewed
by Mayhew in 1849 were quick to point out that their
profit margins had fallen and consequently they were
obliged to work a longer day (Mayhew, 1861, vol. 1,
pp.55, 84). In part this was the result of the
increase mentioned above, but other factors were
also responsible. The sharp fall in food prices
from 1848 to 1852 (Tucker, 1936, p.79; Burnett,
1969, p.198-203) meant that costermongers had to
sell more in order to realise the same profit. A
general fall in prices was paralleled by a rise in
the real wages of artisans. But for the casual
labourer in the sweated trades, wage rates tended
to be reduced, whilst rents in the inner industrial
perimeter actually increased (Stedman-Jones, 1971,
p.216). The urban hawker, therefore, was squeezed
between falling profit margins and increasing rents
and this pressure led to the necessity of working
longer hours.

Judging from the addresses and districts from
which complaints against the 'crying evil' and
'organ nuisance' came, street traders had spread out
from the central districts starting in the early
1850s. Complementing the objections from West End
ratepayers came the suburban voice of protest:
'Quietus' from Islington[17] and 'Pacator' in
Walworth[18] to name but two. The demographic and
financial pressures upon street traders have already
been discussed, and these no doubt were important
factors in the spread of the trade. Faced with
increasing numbers of street sellers and musicians,
the urban bourgeoisie called upon a series of
legislative powers with which to control this
expansion. Three avenues of control and repression
can be singled out as being of particular importance: vestry opposition to street markets as foci
of public disorder; the movement to abolish Sunday
trading; and a more general and diffuse interference with street traders both in the markets and
on the streets. Individually, these moves might
have only been of local significance, but in fact
they complemented each other on a city-wide scale
and at times even merged to form a particularly
cogent legislative arsenal of control and repression.
The central London markets underwent a period

of expansion during the late 1820s and into the
following decade. Smithfield was extended and in
Covent Garden the Duke of Bedford provided new
buildings and took over the control of all market
tolls and stallages (Sheppard, 1971, p.191). New
markets were also planned, such as Hungerford near
the Strand (Fowler 1829) and the South London Market
in Southwark.[19] At first these large markets dealt
in both the wholesale and retail trades but by the
mid-century they were becoming increasingly devoted
to the wholesale sector (Dodd, 1856). In the mean-
time, the unauthorised and informal street markets
which fringed the City borders and served the population of the inner districts, faced displacement
by street improvements. By 1829 St. James' Carnaby
and Westminster markets in the west had been
destroyed (Fowler, 1829, p.7). In 1851 Bloomsbury,
Honey Lane and Oxford markets had all but dis-
appeared (Weale, 1851, pp.611-12). They were
followed a few years later in 1869 by Hungerford
market ('much used' in 1851). Finsbury and Fitzroy
markets, St. George's market in Southwark, Randall's
in Poplar and the People's market in Whitechapel.[20]
Indeed, by 1868 the Corporation of London showed
some concern over the lack of daily markets for its
poor population, as did the authorities in St.
James' Westminster, St. George's in the East,
Rotherhithe and Limehouse.[21]

Street improvements were not the only pressures
brought to bear on the informal marketing system.
From the mid-1840s Evangelical zeal in opposing
Sunday trading, the most important day for working-
class purchases, coupled with parochial opposition,
led to the suppression of a number of street
markets. One of the best-known struggles took
place in 1858-9 between the St. Pancras vestrymen
and the Brill traders in Somers Town. Invoking a
local paving act, Sunday trading legislation and
the 1839 Police Act, or combinations of these, the
vestry made a series of concerted attacks to end
the market. Indeed, on a Sunday the Brill more
resembled a fair than a market: festivities started
at 7 a.m. and spread out into the adjoining streets.
They ceased at 2 p.m. when the Jewish old clothes
sellers returned to Whitechapel.[22] A report from
a local police constable described the scene thus:

> The streets are also thronged with itinerant
> venders of clothes, birds, sponges, pictures,
> fancy baskets and fruits, and indeed almost
> every article that can be mentioned is brought
> here for sale. There are also weighing

140

machines for testing the strength, and on
private ground adjoining the market swinging
boats and roundabouts have been erected. There
is one continual scene of uproar in these
streets; and the yells and howling proceeding
from the market people is beyond description.[23]

Attempts to remove obstructions from the footpath in
November 1858 proved futile. In fact the situation
was made worse with the suppression of the Sunday
markets in Petticoat Lane and Rag Fair when Jewish
traders and their customers transferred to the
Brill.[24] The Metropolitan Police refused to enforce the laws against obstruction as a means of
ending the market, quoting its long ancestry as a
reason for the Brill's continuance. St. Pancras
vestry, however, was not to be deterred and closed
the market in November 1859 under a local paving
act. The Brill was not alone in resembling a fair,
and almost every Sunday market took on that same
appearance. The scenes of disorder and chaos,
with crowds spilling onto the streets when the pubs
closed during the hours of divine service and with
sermons being drowned out by the noise of the
market traders, were repeated every Sunday throughout the city. Consequently local actions to halt
Sunday trading, perhaps significant in themselves
only at a parish level, merged with a more widespread movement to prohibit or regulate street
trading.

In 1850-51 street sellers were harassed and
removed in the City, Southwark and Holborn. In
Southwark shopkeepers used a local paving act to
remove orange sellers from the streets and, under
pressure from ratepayers, police constables were
instructed to prevent obstructions to the footpath
by Hawkers or face dismissal.[25] The Holborn
vestry, spurred on by local shopkeepers, summoned a
number of itinerant fruit vendors but, faced with
an hostile magistrate, failed to secure any con-
victions. The reports of the court proceedings
demonstrated the confusion that existed over the
legal status of street traders and the willingness
of local authority to call upon a variety of
legislative measures in order to suppress them:

Mr. James (Clerk of Holborn Union): "They ought
to walk on the carriage way."

Mr. Combe (Magistrate): "Oh no! The footpave-
ment is free for everybody, and there is no
act of Parliament to compel people to walk on

141

the carriage way."

(An Overseer of the Parish said the calling out
of oranges on Sunday was an intolerable
nuisance.)

Mr. Combe: "They may call oranges all over
the country if they like."

Overseer: "I am an overseer and the inhabitants
expect protection from me."

Mr. Combe: "You are not an act of Parliament:
you are an overseer.   What nuisance is it to
call out 'Two a penny oranges?'"[26]

The regulations concerning street trading and
marketing were indeed confused and no doubt many
dubious actions were brought under the ambiguity of
the law.   Mr. James' use of the term 'carriage way'
was a reference to the Police Act under which
obstructions or annoyance by hawkers in any public
thoroughfare in the Metropolitan Police District
was an offence.[27] Previously the law had been
confined to the City, Westminster and Southwark
only.[28] Local paving acts, as described above,
supplemented this deficiency but at the same time
placed the obligation for enforcement on the parish
rather than on the police.
       Having failed to impress magistrate Combe with
tales of obstruction and annoyance, the Holborn
vestry appealed to his moral conscience as to the
evils of Sunday trading.   No doubt as an educated
man he was quite aware of the interest in the
question, as evidenced by the three select
committees that were set up between 1847-51.
Indeed, the earliest report had drawn attention to
the reluctance of metropolitan magistrates to convict for Sunday trading.[29]   It was quite clear
from the proceedings that one of the main objects
of altering the Sunday trading laws was to call a
halt to street marketing and when amendments to the
existing legislation were proposed they were
soundly criticised as an attack upon the comforts
of the poor.   Three further bills relating to
Sunday trading were proposed between 1848-51, cul-
minating in that described by W.J. Fox as 'a Lambeth
shopkeepers' bill directed against oranges and
apples.'[30] James Hayman, a corn merchant and one
of the promotors of that bill, was quite explicit
as to the desired effects it would have on hawking:

142

The promotors of this bill have considered the subject of hawking, and we think that a greater nuisance does not exist in London than hawking; it has been complained of in every part of the Metropolis and therefore we are anxious to prohibit hawking.[31]

In this fashion the economic self-interest of local ratepayers and the 'shopocracy' drew upon Evangelical zeal to furnish a moral veneer for the control and repression of street markets and hawking. As G.M. Young remarked, 'the virtues of a Christian after the Evangelical model were easily exchangeable with the virtues of a successful merchant or rising manufacturer' (Young, 1977, p.2). Undoubtedly not all Evangelicals considered street trading to be an evil and profane activity as such. But Sunday marketing was an entirely different matter. Observance of the Sabbath was accounted to be crucial to Evangelical discipline, and the fact that Sunday was by far the busiest marketing day for the working class therefore counted for nought. The Sabbath was to be observed and not traded upon. Consequently attempts to suppress Sunday trading did not cease with the defeat of 1851. In 1855 Lord Grosvenor's Sunday Trading Bill and the whole of militant Sabbatarianism clashed with a variety of opposing groups in the Hyde Park trading riots. Although the riots have been shown not to have been a purely spontaneous reaction to the proposed legislation, their duration speaks eloquently for the significance that the issue held for the working class (Harrison, 1965).

Attacks on street trading continued and in 1856 Lambeth vestry ordered the police to rid the streets of any barrows and carts that caused an obstruction in the highways.[32] But it was in the City of London that the strictest measures were enforced. In May 1859 the Court of Aldermen, regretting that 'the evil has increased considerably',[33] instructed the police to clear the streets of hawkers who caused obstructions, annoyed passers-by or impeded traffic during business hours. The police obliged, and in the following year the City costermongers petitioned the authorities to complain about the ill-treatment they were receiving.[34] Herein lies the key to increasing legislative interference with street trading, the Evangelical zeal and parochial opposition would have been impotent were it not for the existence of a city-wide police force able to call upon a series of laws which could be interpreted in ways conducive to such

143

interference.

This had not always been the case, however, as was demonstrated by the reluctance with which the police authorities dealt with the Brill traders. This reluctance to interfere in street marketing stemmed from the conflict it brought not only with the working-class public but also with the magistracy. It had already been made abundantly clear during the battle over the selling of the unstamped press in the early 1830s that the magistrates' sympathies came to rest increasingly with the poor hawkers rather than with the agents of the Commissioner of Stamps (Hollis, 1970). They took a similar view about street traders.[35] Many magistrates were reluctant to convict traders brought before them for contravention of the Sunday trading laws. In 1839 Mayne, then Commissioner of the Metropolitan Police, stated that:

> We have always found magistrates extremely unwilling to commit parties merely for crying for sale, and the facts usually prove to be such as to excite sympathy for the party.[36]

'Sympathy for the party' was in many cases a manifestation of the hostility felt towards the new Metropolitan Police. This hostility took many forms, ranging from a reluctance to pay the police rate to full-scale riots, such as occurred at Spa Fields in 1833 and Hyde Park in 1855. However, our main concern here is with the relationship between the police and street traders.

In the years following the creation of the new Metropolitan force in 1829, the police found themselves hard-pressed to control large crowds or riots. But, as Storch has recently pointed out, the primary role of the constable on the beat was not to preserve public order but to prevent disorder (Storch, 1976, pp.487-96). The importance given to policing the streets bears witness to the constable's role as a 'domestic missionary' whose very presence was intended to instil a sense of urban discipline into the lives and minds of the working class. A constable was expected to patrol his beat steadily and constantly at a rate of two and a half m.p.h., for nine consecutive hours (Reith, 1948, p.32). All beats within a six mile radius of St. Pauls, we are told, were thus traversed in periods which varied from seven to twenty-five minutes, and some points were never free from inspection (Anon, 1852, p.9). In addition, each constable was expected to know the people on his beat not only by sight but also by name, haunts, connections and associates. In conjunction with this knowledge the policeman was instructed to enforce the 'move on' system, and did so with increasing frequency throughout the 1850s. It was in this capacity that they were most likely to come into conflict with the working class in general and street hawkers in particular, whilst at the same time appearing to a middle-class public at their most officious and tyrannical.

If the rate of assaults on policemen can be taken as a surrogate measure of the intensity with which this urban discipline was enforced, it appears to have attained a high level by the late 1850s. Indeed, in 1858 this type of anti-police activity was higher than at any time throughout the following decades of the century (Storch, 1976, p.503). The situation in London before this date is difficult to establish with any degree of precision. Under the 1839 Police Act assaulting a constable was made a separate offence liable to a fine or up to one month's imprisonment, determined summarily at the discretion of the examining magistrate. Before this date it is not possible to distinguish assaults on policemen from common assaults. Nevertheless, comparing the figures for 1842[37] with those of 1859[38], it appears that the number of assaults, obstructions and attempts at rescue almost doubled. Since the figures for the intervening years do not appear to be available, the data does not warrant the drawing of any firm conclusions. Storch, however, believes that the rate of anti-police activity decreased in the late 1850s, although he does not present any conclusive evidence to substantiate this view. If the trend from 1858 is a downward one, then it might be argued that during the period 1842-58, a peak of anti-police activity was reached. But without further evidence a more detailed chronology is impossible and such conclusions as can be reached must remain illustrative rather than substantive.

Despite the shortcomings of the statistical data, it is evident that contemporary opinion was often hostile to police authority. By the time of Mayhew's original interviews with costermongers, to 'serve out a crusher' was already considered to be an act of heroism and he was assured that in the case of a political riot every coster would seize his policeman (Mayhew, 1861, vol. 1, p.20). It was not merely that the police were disliked by the working class alone. Mayne was very conscious of the fact that the sight of a burly 'crusher'

bearing down upon an orange-girl or a waif of a
match-seller did little to endear the force to an
often hostile public opinion.[39] It was noticeable
too, that witnesses were very reluctant to come
forward against street traders, and the City of
London police during their crusade to clear the
streets in 1859 made it a condition of arrest that
persons lodging a complaint should be willing to
appear before the magistrates as a witness. Despite
contemporary opinion, police interference with
street markets and hawkers intensified throughout
the 1850s, but this was not a purely instinctive
reaction on the part of law versus disorder. Rather
it appeared that when local pressures were brought
to bear upon the police they were translated into
concrete action by constables on the beat. This
resulted in stricter enforcement of an urban
discipline designed to prevent disorder occurring on
the streets. In the minds of the authorities,
street markets and traders constituted foci of
disorder and annoyance and as such they tended to
fall foul of this discipline.[40] As a consequence,
police activity in the inner districts was to some
extent responsible for the spread of street trading
into suburban localities. Having been removed
from markets near the centre and forced as a result
to hawk in the public thoroughfares, street traders
were then hindered from carrying out their business
by the 'move-on'system. In effect, legal
marginality was in no small way responsible for the
increasing spatial marginality that came to be a
feature of street trading in the 1850s.

Conclusion

By the end of the 1860s street trading had already
been a marginal occupation for some years. Spatial
and legal marginality were partly the outcome of a
series of attacks upon street marketing and trading.
Financial marginality, intensified by a fall in the
price of food, stemmed from the position in which
costermongers found themselves at the central
markets. It is quite evident, however, that the
original crisis which led to this marginal position
was a rise in the numbers of street traders in the
inner districts. Stemming from this increase,
street traders were obliged to work longer hours and
travel further to hawk their goods or play their
music. In this manner the increasing population of
street traders was in effect produced by a combination of quantitative and qualitative changes.

146

Economic pressures on casual labour were the most profound in structuring the opportunities to gain a livelihood, but they were by no means the only forces in operation. As such it is only by investigating the relationships between urban occupations and the labour market on the one hand, and between the labour market and the political process on the other, that a full understanding of the employment opportunities for casual labour can be acquired.

An understanding of the relationship between occupations and the labour market in which they were embedded is crucial to any wider or comparative studies of nineteenth-century cities. It was precisely because of the unique character of the London labour market that street trading did not appear elsewhere on the scale at which it occurred in the capital, but rather than being dismissed as an isolated phenomenon with no wider significance, street trading has been shown to require critical historical analysis. Whilst many occupations in the nineteenth century remained peculiar to particular cities or regions, all were in some way related to local labour markets. Without investigating this relationship, or at least being aware of its implications, no sense can be made of the uniqueness of many urban occupations. Since it might be more correct to state that there was no archetypal 'nineteenth-century city' - only a variety of cities - the question of typicality is an important one. What seems clear is that the very uniqueness of particular urban occupations may often be used to gain insights into the more general economic structure of nineteenth-century cities.

Notes

1. Parliamentary Paper (PP.) 1852-3, LXXXVIII. Decennial census...for London, p.28
2. PP. 1888, LIII. Royal Commission on Market Rights and Tolls: minutes of evidence, p.169
3. Bebbington, J. (1882). Autobiography, folio 357. Solly Manuscript Collection, British Library of Economic and Political Sciences
4. PP. 1888, LIII. R.C. on Market Rights......, p.166-7
5. *ibid.*, p.174
6. Weekly Despatch, 24th November, 1839
7. Public Record Office. Inland Revenue Papers, IR 51/5. Petititon from David Hammond, 24th April, 1832
8. Friendly Association of London Costermongers instituted 23rd June, 1850, p.3. Goldsmiths Library, London
9. Contemporary references to St. Giles were numerous

but amongst the most influential were 'The state of the inhabitants and their dwellings in Church Lane, St. Giles', *Journal of the Statistical Society*, XI., and Beames, T. (1850) *The Rookeries of London*, Bosworth, London

10. *The Times*, 1st January, 1845
11. *ibid.*, 2nd May, 1856
12. *ibid.*, 10th November, 1857
13. *ibid.*
14. *ibid.*, 23rd January, 1858
15. *ibid.*, 15th April, 1858
16. 27 and 28 Vict. c.55
17. *The Times*, 20th July, 1852
18. *ibid.*, 11th November 1857
19. South London Market Company, (1832) *Prospectus*. Broadside Collection, Guildhall Library, London. Though planned the market appears never to have been built.
20. Guildhall Record Office, London. Report to the Court of Common Council from the Markets Committee 16th December, 1869. Co.Co. Printed Minutes 1869, App. 31
21. Guildhall Record Office. Returns to questionnaires submitted to municipal authorities by F. Brad, Comptroller, on behalf of the Corporation requesting information on markets...1868. Co.Co. Misc.Ms. 314
22. Camden Local History Library. Heal Collection, *Illustrated News of the World*, 22 May, 1858
23. Public Record Office, Metropolitan Police Papers, MEPO 2/47. Somers Town Market: complaints against Sunday traders, obstructions of foot and carriage ways and possible reports, 2nd December, 1858
24. *ibid.*, letter to Sir Richard Mayne from Thomas Watts, vestry clerk to St. Pancras, 2nd May, 1859
25. *Reynolds Weekly Newspaper*, 6th October, 1850; *Daily News*, 14th January, 1851
26. *Reynolds Weekly Newspaper*, 26th January, 1851
27. An act for further improving the police in and near the Metropolis (The Police Act), 2 and 3 Vict. c.47
28. An act for better paving, improving and regulating the streets of the Metropolis and removing and preventing nuisances and obstructions therein (Michael Angelo Taylor's act), 57 Geo.III. c29
29. PP. 1847, IX. Report from the Select Committee on Sunday Trading (Metropolis), p.iv.
30. *Hansard*, 24th July, 1850, p.201
31. PP. 1850, XIX. Report from the Select Committee of the House of Lords appointed to consider the bill intituled "An act to prevent unnecessary trading on Sunday in the Metropolis": minutes of evidence, p.58
32. Lambeth Local History Library. St. Mary's (Lambeth) vestry minutes, 22nd August, 1856
33. *City Press*, 7th May and 14th May, 1859
34. *The Times*, 23rd November, 1860
35. *Weekly Despatch*, 2nd June and 18th August, 1839

36. P.R.O. Metropolitan Police Papers, MEPO 1/45.
Commissioner's letter book 1836-1850. Letter from
Sir Richard Mayne to Sir William Coutts, 29th November, 1839
37. PP. 1844, XLVI. Metropolitan police returns 1842,
p.157
38. PP. 1860, LXIV. Judicial statistics. Table 7,
Offences determined summarily...1853
39. PP. 1850, XIX. Report from the select committee...
to consider a bill intituled "An act to prevent unnecessary
trading...": minutes of evidence, p.122, 125
40. In an essay on the social control of working-class
leisure, Hugh Cunningham has stated that inner London was
left bereft of fairs by the 1850s. This should be qualified
by noting that Sunday markets such as occurred in the New
Cut, the Brill and Whitecross street, were as much fairs as
they were markets. See Cunningham, H.(1978) 'The
Metropolitan fairs: a case study in the social control of
leisure', in A.P. Donajgrodzki (ed.), *Social Control in
Nineteenth Century Britain*, Croom Helm, London, pp.163-84

References

Alexander, S. (1976) 'Women's work in nineteenth century
London: a study of the years 1820-50.' in J. Mitchell and
A. Oakley (eds.) *Rights and Wrongs of Women*, Penguin,
Harmondsworth, pp.59-111
Anon (1823) *The new cries of London or itinerant trades
of the British Metropolis*, No publisher, London
Anon (1852) 'The Police system of London', *Edinburgh
Review*, 195, 1-33
Archer, T. (1865) *The pauper, the thief and the convict*,
Groombridge and Sons, London
Babbage, C. (1864) *A chapter on street nuisances*, John
Murray, London
Bass, M.T. (1864) *Street music in the Metropolis*, John
Murray, London
Beveridge, W.H. (1909) *Unemployment - a problem of
industry*, Longmans, London
Burn, J.D. (1855) *The autobiography of a beggar boy*,
William Tweedie, London
Burn, J.D. (1858) *Commercial enterprise and social
progress*, Stephenson and Spence, London
Burnett, J. (1969) *A history of the cost of living*,
Penguin, Harmondsworth
Dodd, G. (1856) *The Food of London*, No publisher, London
Dyos, H.J. and Aldcroft, D.H. (1974) *British transport:
an economic survey from the 17th century to the 20th*, Penguin,
Harmondsworth
Engels, F. (1973) *The condition of the working-class in
England*, Progress, Moscow. (First English translation 1892)
Fowler, C. (1829) *Description of the plan for the*

revival of Hungerford market, No publisher, London

Hall, P. (1962) *The Industries of London since 1861,* Hutchinson, London

Harrison, B. (1965) 'The Sunday trading riots of 1855', *Historical Journal,* 8, 219-45

Hobsbawm, E. (1964) 'The nineteenth century London labour market', in R. Glass (ed.), *London: aspects of change,* Macgibbon and Kee, London

Hollis, P. (1970) *The Pauper Press,* Oxford University Press, London

Kirwan, D. (1963) *Palace and hovel,* Abelard - Schuman, London

Lees, L. (1976) 'Mid-Victorian migration and the Irish family economy', *Victorian Studies,* 20, 25-43

Leighton, J. (1851) *The London cries and public edifices,* Griffith and Farran, London

Mayhew, H. (1861) *London labour and the London poor,* Griffin Bohn, London

Murray, J.F. (1843) *The world of London,* Blackwood, London

Price-Williams, R. (1885) 'The population of London 1801-1881' *Journal of the Statistical Society,* 58, 349-432

Reith, C. (1948) *A short history of the British police,* Oxford University Press, London

Rudé, G. (1964) *The crowd in history,* John Wiley, New York

Rudé, G. (1974) *Paris and London in the eighteenth century,* Fontana, London

'Sam Syntax' (1820) *Sam Syntax's description of the cries of London,* No publisher, London

Sheppard, F. (1970) *London: the infernal wen 1808-70,* Secker and Warburg, London

Smith, C.M. (1857) *The little world of London,* Hall, Virtue and Co., London

Smith, J.T. (1817) *Vagabondiana: or anecdotes of mendicant wanderers through the streets of London,* No publisher, London

Smith, J.T. (1839) *The cries of London,* No publisher, London

Stedman-Jones, G. (1971) *Outcast London,* Oxford University Press, London

Stern, W. (1966) 'The Baronness's market: the history of a noble failure', *Guildhall Miscellany,* 2, 353-66

Storch, R. (1976) 'The policeman as domestic missionary: urban discipline and popular culture in northern England 1850-80', *Journal of Social History,* 4, 481-509

Thompson, E.P. (1967) 'The political education of Henry Mayhew', *Victorian Studies,* 11, 41-62

Thompson, E.P. and Yeo, E. (1973) *The unknown Mayhew: selections from the Morning Chronicle 1849-50,* Penguin, Harmondsworth

Tucker, R.S. (1936) 'Real wages of artisans in London

1729-1935', *Journal of the American Statistical Association*, 31, 73-84

Ward, D. (1976) 'The Victorian slum: an enduring myth?', *Annals of the Association of American Geographers*, 66, 323-36

Weale, J. (1851) *London and its vicinity exhibited in 1851*, John Weale, London

Young, G.M. (1977) *Portrait of an age*, Oxford University Press, London

The research for this article was made possible by an S.S.R.C. studentship.   The author would also like to thank Jeff and Carol Johnson for their encouragement during the period of research.

Chapter 6

RETAILING IN THE NINETEENTH-CENTURY TOWN : SOME PROBLEMS AND POSSIBILITIES

Roger Scola

Background

It might seem a trifle churlish to begin a chapter
on a subject that has been allotted a whole section
in this collection of essays by remarking that it
is a neglected aspect of economic and urban history.
However, the service sector in general has always
come a poor second to studies of the changes in the
primary and manufacturing industries, and retailing
can feel particularly hard done by.   Retailing was
an important industry in its own right; in many
towns it must have provided a major source of
employment, and in the markets and shops tied up a
considerable amount of fixed capital.   Yet the
subject but flits across the pages of the general
economic history textbooks and the state of our
knowledge can be summed up briefly.
    There has developed a division of opinion
between those espousing the view expressed by
Jeffreys (1954) that retailing remained an under-
developed part of the economy until certainly the
second half of the nineteenth century, and those
following Clapham (1926,' volume I, Chapter VI) who
suggested that there had already been important
changes before 1850.   It would not be correct to
see this as a controversy as such; the subject was
too low key for that; and in any case, Jeffreys'
work was impressive enough to carry the day.   His
general view that retailing bore the marks of a
pre-industrial economy also received strong backing
from Dorothy Davis (1966, p.252).

> '...it seems as though so much of the nation's
> energy and imagination was being drawn into
> industry and commerce that none was left to
> fertilise the retail trades with new ideas.
> Perhaps a dull calling attracted dull recruits.'

From Mrs. Davis and others, it was not until the last quarter of the century that the dull and dowdy retailer emerged from his chrysalis as the glittering butterfly that we know today (Mathias, 1967; Yamey, 1954). Such was the transformation, indeed, that he was summoned forward to represent the spirit of dynamic entrepreneurship so sadly lost by others on the playing fields of Eton (Wilson, 1965). Thus if there was anything of interest to be found in the history of retailing, it was to this later period that we should look.

As Jeffreys and Davis saw it, for much of the nineteenth century, consumer goods were either made or grown at home or bought from the producer-retailer operating from his workshop or perhaps from a market. There were some ordinary retailers, most noticeably in the haberdashery and drapery trade, but the main thrust of their argument is that the bulk of fixed shops had a middle class clientele with the working classes frequenting markets and itinerant traders. Essentially, therefore, later nineteenth-century developments in urban retailing could be seen, for the middle classes, as an increase in the number and size of retailers with particular emphasis on the growing gap between producer and seller, and for the working classes, as a movement away from markets and the like with the transference of trade to shops as we know them today.

This view, however, has been modified in two important ways. Firstly, Janet Blackman (1963) suggested that markets and shops were not competitors for working-class custom but complimentary, with the purpose of the former being primarily to supply perishable foods such as meat, vegetables, and fruit. Secondly, David Alexander (1970) demonstrated that shops existed in the first half of the nineteenth century in much larger numbers than had been thought. My own work on Manchester has lent support to both these criticisms, and in fact, suggests that the rise of the small general food shop dated from the later eighteenth century, and rapidly became central to working-class life in large industrial towns (Scola, 1975).

The central proposition of Jeffreys' and Davis' thesis that nineteenth-century retailing was largely the story of the weaning of the working classes away from street traders into shops seems highly questionable. However, it is not just the changes in the structure and chronology of retailing in the nineteenth century that need reappraisal, for we also need to consider the specific impact of

154

towns upon these processes.   Recent writing on
retailing brings us face to face with the
perennial nightmare of all those interested in
urban studies: were the developments in retailing in
the nineteenth century distinctively urban, in contrast to what happened in the countryside, or did
they just occur first in towns because most people
came to live in towns?   Both Blackman (1976) and
Alexander (1970) seem to place the emergence of the
small general shop quite explicitly in the context
of the growth of a non-food producing urban population, but it is a theme implicit in most of the
writing on the subject.   Yet the distinctively
urban quality of fixed retailing has never been
entirely convincing, since the village shop has been
seen as a traditional feature of rural life, and
there can now be little doubt that such outlets
existed in considerable numbers in the seventeenth
and eighteenth centuries, if not even earlier.   Nor
were they just to be found in market towns as we
might expect, but also in smaller towns and villages
and comprising a wide range of retailers including
butchers, bakers, ironmongers, haberdashers, and
shoemakers (Millan, 1976; Brown, 1969).

Sources

One might be forgiven for wondering why it has
taken so long to answer that most basic question of
all: did shops actually exist?   Mainly, it reflects
the fact that the question was rarely asked, but
the difficulties of the potential sources compound
the problem.   The most obvious sources are the
directories that date from the later eighteenth
century (Norton, 1950), but there are considerable
problems in their use.   Commercial directories
require the disentangling of multiple entries, and
of retailers from wholesalers, but much more
serious in those compiled before the 1840s is the
limited coverage given to small shopkeepers and
market traders.   These omissions and their subsequent inclusion as the directories become more
comprehensive, do tend to exaggerate the rate of
growth of retailing in the later period.   Early
unclassified directories are rather better in this
respect, although the obvious step of cross-checking
directory entries with rate books to ensure that
the listings do represent retail outlets is unhelp-
ful since many small shops were not listed for
rateable purposes (Davies, Giggs, and Herbert,
1969).   Where rate books have been useful in my

own work, however, is in providing information about the occupancy of market sites and this information may also be available for other large towns. Before the later eighteenth century the problem is more difficult. A range of sources have been used to throw some light on the existence of retailers, including the evidence of trade tokens, lists of freemen, guild records, probate inventories; but all are inevitably fragmentary and selective, failing to match up to even the poorest of the directories (Willan, 1976; Brown, 1969; Borsay, 1977).

The urban market

Although I have strayed beyond the limits of the nineteenth century in discussing sources, this wider time span is essential because we cannot appreciate what was distinctive about retailing in the nine-teenth century unless we know what went before. This is particularly well demonstrated by looking at the role of the market in urban retailing. Dorothy Davis and others suggest that the purchase of goods at the market was a stage through which the urban consumer went on his way to reliance on the corner shop. Shopping at a market was a habit that the migrant brought from the countryside and one that he clung to in an alien environment. Alexander talks in terms of a cultural gap between the retailer and his customer which was more easily bridged in the informal atmosphere of the market place, and it might even be suggested that the con-tinued importance of urban markets was a measure of the adjustment the town dweller had made to his new surroundings. A variation of this argument is that markets had an emotional appeal, based on the idea of fair play incorporated in what has been called 'the moral economy of the crowd' (Thompson, 1971). Rudé (1967, p.90) has highlighted the 'ideology of the old English village' that included a belief in a fair wage and a just price, and markets could be said to embody these ideals as places where producer and consumer could meet on equal terms.

Both these general views of the reasons for the importance of urban markets must now be suspect. Evidence of widespread rural shopkeeping before the nineteenth century must make us doubt whether the new town dwellers were so unused to buying their needs at shops that they naturally looked to the market place, or that they should hold the latter in such awe. In fact, the function of the urban

market in the distribution system was a good deal
more complex than this.   In the first place, not
all markets prospered.   Outlying markets in
Manchester and Salford, ostensibly in the best
situations nearest to the fastest growing parts of
the town, met with very limited success, and a
similar situation has also been documented in Leeds
(Scola, 1975; Grady, 1976).   It was the markets in
the centres of towns that seemed to succeed best.
Secondly, it is clear that, in the case of
Manchester, the retail markets were not populated
by producer-retailers even at the beginning of the
century, but by a mixture of traders, and that the
composition of traders in the markets changed over
time.   Butchers were ordinary retailers increasingly anxious from the 1830s to leave the market
sites; fishmongers similarly were ordinary retailers,
but even in the 1870s and later were still content
to remain there; the sellers of vegetables and
fruit came nearest to the traditional image of
market traders, as producer-retailers maintained an
importance here until mid-century.   By then, most
markets had come to include a variety of other
traders such as the sellers of second-hand clothing,
pots and pans, trinkets, hot pies and saveloys, and
their place in the food distribution system had
changed.   They were increasingly concerned with
wholesaling rather than retailing and their
importance for many, as real incomes rose and
leisure time increased, was to provide an opportunity for a Saturday outing and the chance to supplement a monotonous diet with some seasonal vegetable
or fruit.
       Markets did, at least, generate historical
records.   They occupied central sites; they spilled
out into the highway and generally disrupted the
life of the surrounding areas; market authorities
jealously guarded their rights and often clashed
with their tenants; the building of covered market
halls were matters of civic pride.   These
activities raise many interesting themes for urban
historians, but they must not divert us from
looking at the primary function of urban markets.
My research has been based on a large industrial
city (Manchester) and the chronology of changes in
markets may well be different in other sorts of
towns.   From the trail of assorted legal and
administrative records, however, it should be
possible to build up a picture of the types of
traders operating from market sites in any town, of
the reasons for their presence, and the part that
they played in supplying the needs of the urban

population.

The structure of urban retail provision

Several studies of markets have pointed to a surprisingly large number of butchers in nineteenth-
century towns (Blackman, 1963; Alexander, 1970;
Scola, 1975). The consumption of meat, of course,
is one of the key issues in the debate over the
standard of living of the urban worker in the first
half of the century (Taylor, 1975), and the presence
of butchers on this scale in industrial towns would
seem to have some implications for the controversy.
The connection between provision of retail
facilities and levels of consumption has been used
in a general way for earlier centuries (Fisher,
1934-5; John, 1961), but it may be applied more
specifically to nineteenth-century towns. One
obvious way is to use directories to look at the
number of retail outlets per head of population in
different places. Alexander (1970) examines eleven
towns in this way and shows that the level of
retail provision improves over the period 1820-1850,
but that, for instance, it varies considerably from
the lavish one shop to every 50 people in Nottingham
to the positively meagre one to every 145 in
Merthyr Tydfil in 1850-1. The differences are
striking and easy to discern, but there are considerable drawbacks to this analysis. The
accuracy of directories will vary (as described
above) both from place to place and over time, and
while this can be overcome, it requires a good deal
of careful work on each town to achieve an acceptable level of consistency. Furthermore, the use
of urban populations at different times to represent
the catchment area for the various shops and stalls
is likely to produce some very distorted results.

This particular aspect is worthy of more
attention. Many towns served as centres for the
surrounding countryside and other small towns and
villages in the region. Consequently they would
have contained a much higher number of retailers
than would have been strictly warranted by the
immediate urban population. Conversely, some
towns appeared to have relatively poor retail
facilities unless those in near-by larger towns are
also taken into account. This was less true,
perhaps, for food retailers than for others, but it
may still apply to sellers of more expensive foodstuffs that were only bought infrequently.

Hence, it makes little sense to examine the

level of retail facilities in any one community
without reference to that town's relationship with
the other towns and villages in the area, and to the
way that this relationship changed over time. A good
example of this is provided by Nottingham. Taking
Alexander's figures at face value, Nottingham had a
very favourable shop-population ratio and one that
improved faster than most before 1850. These
statistics must, however, be considered in the light
of the particular pattern of urban development.
Nottingham comprised a compact central town surr-
ounded by a number of satellite industrial towns and
villages that only later became integrated into the
city. As a result, the true catchment area for
Nottingham's retailers was a good deal larger than
the population of the central city. Gradually,
these surrounding towns established a wider range of
shops, especially for foodstuffs, while the city
itself remained the natural location for higher level
retailers. Indeed, in the period 1835-50 (when
data is more reliable), the increase in the number
of shopkeepers, grocers, and bakers in central
Nottingham barely keeps pace with population and
lags behind that of other towns, while the improvement in the ratio of non-food shops per head of
population is almost six times as great as for food
retailers. It is this kind of interdependence in
the retail facilities of near-by towns that Wild and
Shaw (1975) have demonstrated in the Calder valley
during the second half of the century, as retailers
responded not just to population growth and changes
in levels of income but also to increased mobility.
Realization of the extent of fixed retailing before
1850 should encourage others to push back this sort
of analysis even into the eighteenth century.

Despite these problems of interpretation, the
technique of calculating shop-population indices
would still seem to be worthwhile, if applied
selectively. Food retailers are particularly
attractive in this respect. The area from which
they would expect to draw their custom was relatively
compact and they would suffer from fewer of the
complications outlined above than other shopkeepers.
Certainly, in Manchester and Salford, such calculations produce some interesting results (Table 6.1).

To attribute these variations in shop-
population indices to under-recording or an underestimated catchment area would require that either
the retail outlets are under-recorded by something
like a third in the 1820s and 30s, or that the true
catchment area in 1800-11 was a third greater than
the city itself. Both suggestions seem unlikely,

159

Table 6.1: Population per food retail outlet

1800	1811	1831	1841	1851	1861	1871
119	113	148	146	121	97	83

and in fact the rapid growth of Manchester and
Salford in the 1820s and 30s appeared to outstrip
the retail facilities of the town. The quality of
life must have worsened in so far as shops and
markets became much more crowded and there must have
been a significant shift in economic power away from
the consumer towards the retailer, affecting both
the quality of service and of the food sold. It was
only gradually, and most noticeably during the 1840s,
that the level of retail demand within the town
seemed sufficiently buoyant to encourage new
retailers to enter the trade and so restore the
shop-population ratio to that of 1800-11.

Such comparisons of the level of retail provision, do, of course, make the assumption that the
average size of outlet remained constant. Super-
ficially similar figures could well disguise important
differences. The typical draper in a city centre
or a prosperous market town would almost certainly
be larger than the typical draper in a small
industrial town. The problem in one sense is
insoluble. There are only sporadic records that
give levels of turnover for particular traders
(Alexander makes particularly good use of the
records of bankrupt shop keepers), and there is no
way of assessing the typicality of such documents.

This fragmentary data may produce acceptable
results for a limited range of retailers in
particular areas, but comparisons of a broader sort
can only be made with a more systematic data source.
The occupational censuses from 1841 provide some
information in this respect, and it is a temptingly
straightforward exercise to link census returns
with the number of outlets, to arrive at some indication of the relative average size of different
sorts of retailers in different towns. The main
problem, however, is that we have to distinguish in
the census data between wholesalers and retailers,
and between those producing and processing the
product and those actually selling it. Only
detailed studies of particular trades can help to
sort out these ambiguities.

Perhaps the most interesting use of occupational statistics is not just to investigate the

160

level of retail facilities at different times, but
to indicate the varying patterns of retail structure
to be found in different towns.   Economic
historians often pay lip-service to the need to know
more about regional differences, and nowhere more
so than in the quest for the answer to the 'standard
of living' debate, but there are surprisingly few
studies which specifically pursue their theme in
this way.   The study of retailing may be significant here, as differences in the way that people
spent their wages should be reflected in the varying
importance of different sorts of retailer.   Tables
6.2 and 6.3 attempt to construct a profile of the
food traders in five, selected towns using data
taken from Alexanders calculation of retail outlets
(1970, Appendix I), my own calculations for
Manchester, and the published census occupational
figures.

Table 6.2: Outlets expressed as a percentage of total (1848-51)

	Manchester and Salford	Leeds	Nottingham	Norwich	York
shopkeeper	46	50	40	36	44
grocer & tea dealer	10	8	8	16	11
baker	8	3	15	24	8
confectioner	4	4	3	4	4
flour dealer	3	8	-	2	2
provision dealer	1	3	3	-	-
butcher	23	20	28	11	24
fishmonger	1	1	3	4	2
poulterer	1	1	-	-	2
greengrocer	3	2	-	3	3
	100	100	100	100	100

A number of points can be made from these
tables.   First, the difficulty of linking census
data with the number of outlets is apparent.   This

is partly a problem of classification as 'shop-keepers' in the directories emerge as 'grocers' in the census, but it also reflects a genuine confusion over where the dividing line, if any, could be drawn between the first six trades listed (Scola, 1975). Thus, for instance, only the census figures bring out the importance of Yorkshire home-baking (Allen, 1968).

Table 6.3: Census occupations expressed as a percentage (1851)

	Manchester and Salford	Leeds	Nottingham	Norwich	York
shopkeeper	17	26	10	15	9
grocer & tea dealer	20	22	24	29	21
baker	21	7	20	23	8
confectioner	7	7	8	7	28
flour dealer	1	2	-	-	1
provision dealer	1	-	1	-	-
butcher	19	27	26	14	25
fishmonger	4	3	7	7	3
poulterer	1	-	-	1	1
greengrocer	9	6	4	4	4
	100	100	100	100	100

Further interpretation of the figures is diffcult. We need to know more about the relationship of market towns such as York to their surrounding areas, but the contrast in the importance of butchers in York and Norwich is sufficiently large to suggest that the consumption of meat was a good deal higher in the former town. Similarly, the proportion of butchers to be found in Manchester is mirrored

162

often enough in other Lancashire towns to suggest
that is is some guide to levels of regional meat
consumption. Likewise, the fact that in both sets
of data, Norwich had one fishmonger to every two or
three butchers, in contrast to the smaller proportions of fishmongers in Leeds and Manchester, has
obvious implications for the importance of fish in
the diet.

It must also be noted that a comparison of the
two tables seems to show a tendency to under-record
the number of outlets for fishmongers and greengrocers, particularly in the two largest towns.
The existence of retailers (from the census) without
any obvious trading outlet, leads us to one of the
most difficult questions in any study of urban
retailing: did itinerant traders play an important
part in the distribution of urban foodstuffs?
Market retailers without regular tenancies and the
various sorts of hawkers are difficult to take into
account in any assesement of retail facilities, and
indeed, might be thought to undermine the already
shaky foundations of much of the preceding analysis.
However, discrepancies between the two tables may
in part be explained by the fact that some outlets
undoubtedly employed more people than others and
some from the census lists worked for wholesalers.
Both explanations apply to fishmongers. In the
case of greengrocers, it is because they did not
occupy fixed sites in the markets, and hence
escaped the compilers of directories and rate-books.
What these figures do not indicate, however, is
that hawkers played an important role in large industrial towns such as Manchester.

Much of the research on hawkers and casual
traders is inevitably coloured by Mayhew's graphic
account of London (Mayhew, 1861-2, volume I), but
in this respect, as in many others, London is
exceptional. It had very large wholesale markets
established well before the nineteenth century, and
sprawling suburbs long before the real incomes of
the working classes had risen high enough for high-
quality foods to be consumed in any quantity. The
natural response to these circumstances was the
creation of a standing army of casual sellers,
flexible enough to switch between the different
commodities available, and mobile enough to cover a
sufficiently large catchment area in and around
London to eke out a living. In most large
industrial towns of the early-nineteenth century,
however, there existed neither wholesale markets
nor suburban development on this scale. Demand for
high-quality foods could thus be catered for by

163

producer-retailers or by established market retailers, with hawkers playing a much more peripheral role in food distribution than in the capital.

The locational behaviour of retailers

It has been suggested that market traders varied in their willingness to stay in the market or to open shops, and that such decisions must have been closely related to the level of demand for their product. This analysis can be carried a little further through a more systematic study of the locational behaviour of retailers. Wild and Shaw (1974) have demonstrated the utility of commercial directories in understanding the geography of retailing in nineteenth-century towns, but this sort of analysis also has much to offer economic historians. The literature on location theory is voluminous, but the part most appropriate to the historical study of urban retailing is the concept of demand threshold. At its simplest, central place theory is based on two premises; first, that any tertiary activity will be located in the place that maximizes the potential market area; and, second, that all things being equal, the consumer will utilize his nearest outlet. These two premises are not immediately compatible, in that while retailers will gain the largest market area by a central location, the customer may be reluctant to travel any further than necessary to make a purchase. The concept of a demand threshold seeks to rationalize this conflict by suggesting that, assuming retailers require similar levels of turnover or profitability, the larger the market area required to attain that level, the greater will be restrictions on suitable locations and the greater the tendency for retailers to stay in the centre of the town. A number of factors will clearly confuse the simplicity of this theory, including variations in the sizes of outlets, the cost of sites and the ability of any shop to attract more distant custom by attractive service or prices, but if we confine our study to food retailers before 1870, many of these complicating factors are of much less importance than in the last quarter of the century when competition became more overt. In short, we may assume that the overriding factor influencing food retailers' behaviour was the size of the catchment area necessary to ensure an adequate level of turnover and this, in turn, must have reflected the level of consumption of particular foods.

To examine this hypothesis, I studied the locations of six different food retailers (shopkeepers, grocers and tea dealers, bakers, butchers, fishmongers and greengrocers) from the directories and rate books of 1811, 1850, and 1871 for Manchester and Salford. First, I assessed the extent to which retailers clung to the centre of the city, and the rate at which they developed in the growing suburbs of the towns. Throughout the period, shopkeepers were the retailing pioneers setting themselves up most quickly in new areas and in 1811, the locational behaviour of shopkeepers was in sharp contrast with all trades except bakers. By 1850 and especially by 1871, however, grocers also had begun to disperse to peripheral locations. Most noticeable was the speed with which butchers, once relieved in 1846 of manorial restrictions to trade from the markets, located themselves in the newest parts of town. By 1871, their locational pattern was closest to that of shopkeepers. On the other hand, fishmongers and greengrocers were much slower to desert central sites with six out of ten outlets in the town centre in 1850, and a third even in 1871.

It is also possible to see differences in the sort of location that these retailers chose as they set up their businesses away from the centre. Business addresses were divided into three categories: major main roads; minor main roads; other roads or streets; and detailed differences in locational behaviour analysed. Once again, shopkeepers were first to locate away from the main thoroughfares, and by 1811 the small corner shop in the back street was already common, with two-thirds of shopkeepers located in the third category of street. This behaviour was strikingly different to that of all other retailers. Even in 1850, two-thirds of the others were to be found in main road locations, except for butchers who were evenly split between the categories. By 1870, shopkeepers and butchers had been joined by fruiterers with most of their shops in side streets. Of the others, fishmongers were most likely to be found in main road positions. It would seem reasonable to argue that main road sites maximized potential market areas, and that only retailers who were very sure of being able to generate enough custom from the inevitably smaller catchment area that back streets would have offered, would locate away from a main road.

A further question that may be investigated relates to whether retailers spread evenly through the different parts of the town, or whether they congregated in certain areas. If we included a wider range of retailers and not just food-retailers, we might expect to see the development of secondary shopping centres in suburban locations, but we may also investigate the extent to which the range of food shops varied between areas of different social composition. We might hypothesize, for example, that butchers clustered in better class areas and that this explains their relatively large numbers. The most straightforward way of describing the retail structure of different areas is to use the local location quotient (Hall, 1962; Lee, 1971). It is calculated by dividing the number of outlets of, for example, grocers in a particular area, expressed as a percentage of all grocers in the town, by the number of all retailers (in this case, the group of six food retailers already referred to) in that area, expressed as a percentage of the overall total. If an area has only 5% of grocers but 10% of all food retailers, then the local location quotient of grocers for that area is 0.5; if there are more grocers in an area than expected given the overall level of retail provision, then the quotient is greater than 1.

A map of Manchester and Salford in 1870 was divided into 34 areas. The local location quotient was then calculated for the six food retailers in 1811, 1850 and 1871 for each area. By 1850 every area had some shopkeepers, bakers, and grocers, and very few areas had a marked under- or over-representation of these trades (only 17 out of 102 L.L.Q. values were outside the range 0.5 - 1.5). More surprising was that butchers were also evenly distributed in 1850 with some outlets in every area, and in only three areas were they noticeably under-represented with a L.L.Q. of less than 0.5. Quite clearly, the large numbers of butchers listed in the directories were not concentrated in a few upper class areas. The contrast with fishmongers and greengrocers is, however, quite marked. These were absent or poorly represented in many parts of the city in 1850, and those areas outside of the centre where fishmongers and greengrocers were well represented included some of the most desirable residential areas of Manchester and Salford. By 1871, however, even these two trades showed the same even distribution. Greengrocers were in every area and had very few high or low L.L.Q.s. Fishmongers still had a propensity to cluster in the

166

better parts of the suburbs, but their under-representation in other areas was much less obvious than at the earlier date.

This sort of locational analysis not only describes the pattern of nineteenth-century retailing but may also offer valuable insights into the consumption of different foodstuffs at different times and in different areas. However, one interpretation of the even distribution of food retailers is that few areas were markedly inferior or superior in social and economic terms. It could be argued, for instance, that differences in the social composition of the various areas were too slight to be reflected in a crude analysis of retail provisions. Manchester was often said to be a predominantly working-class city, deserted by the middle class (Kay, 1832; Engels, 1845), and the similarity of the location quotients could be reflecting this supposed homogeneity.

Yet the argument is not totally convincing. There were observable differences between parts of the city, but the retail provision of generally acknowledged middle-class areas in the outer suburbs was not noticeably better, and in some instances worse, than in some of the least respectable districts. Deficiencies in the directories may account for some of these discrepancies, but there are two other explanations. Firstly, partially overriding the influence of social composition and income level in any one area, was the reluctance of higher order retailers to leave a central site or busy main road even in a working-class district. The second explanation follows from this and relates to the assumption that customers were unwilling to travel long distances to a retail outlet. Willingness to travel will reflect comparative costs, the time available and the effort required for such a journey. That the better-off were less sensitive to higher costs is obvious, but time and effort were also of less significance for this group, since it was not their time and effort that were being expended. Servants could travel to central shops, or the household could afford to deal with those retailers who collected orders and made deliveries, thus reducing the need for shops in high-class suburbs.

While there are obvious pitfalls in this analysis, we have made some progress from the rather simplistic picture of retailing changes in the nineteenth-century town with which we began, towards a better understanding of the complex and often confusing processes at work. The merging of economic history and geography can make an important contribution in this respect by greatly strengthening the analytical tools available to us and by allowing new areas of investigation to be developed.

References

Allen, D.E. (1968) *British tastes*, Hutchinson, London

Alexander, D. (1970) *Retailing in England during the Industrial Revolution*, Athlone Press, London

Blackman, J. (1963) 'The food supply of an industrial town: a study of Sheffield's public markets 1780-1900', *Business History*, 5, 83-97

Blackman, J. (1976) 'The corner shop: the development of the grocery and general provisions trade', in D.J. Oddy and D.S. Miller (eds.), *The making of the modern British diet*, Croom Helm, London, pp.148-60

Borsay, P. (1977) 'The English urban renaissance: the development of provincial urban culture c.1680 - c.1760', *Social History*, 5, 581-603

Brown, A.F.J. (1969) *Essex at Work, 1700-1815*, Essex County Council, Chelmsford

Clapham, J.H. (1926) *An economic history of modern Britain*, Cambridge University Press, Cambridge

Davies, W.K.D., Giggs, J.A., and Herbert, D.T. (1969) 'Directories, rate books and the commercial structure of towns', *Geography*, 53, 41-54

Davis, D. (1966) *A history of shopping*, Routledge and Kegan Paul, London

Engels, F. (1845, 1958 edition) *The condition of the working class in England*, Blackwell, Oxford

Fisher, F.J. (1934-5) 'The development of the London food market, 1540-1640', *Economic History Review*, First Series 5, 46-64

Grady, K. (1976) 'Profit, property interests and public spirit: the provision of markets and commercial amenities in Leeds, 1822-29', *Publications of the Thoresby Society*, 59, 165-95

Hall, P.G. (1962) *The industries of London since 1861*, Hutchinson, London

Jefferys, J.B. (1954) *Retail trading in Britain 1850-1950*, Cambridge University Press, Cambridge

John, A.H. (1961) 'Aspects of English economic growth in the first half of the eighteenth century', *Economica*, New Series 28, 176-190

Kay, J.P. (1832, 1969 edition) *The moral and physical condition of the working classes employed in the cotton manufacture in Manchester*, E.J. Morten, Manchester

Lee, C.H. (1971) *Regional economic growth in the United Kingdom since the 1880's*, McGraw-Hill, London

Mathias, P. (1967) *Retailing revolution*, Longmans, London

Mayhew, H. (1861-2) *London labour and the London poor,* Griffin, Bohn & Company, London

Norton, J.E. (1950) *Guide to the national and provincial directories of England and Wales...published before 1856,* Royal Historical Society, London

Rudé, G. (1967) 'English rural and urban disturbances on the eve of the first reform bill, 1830-31, *Past and Present,* 37, 87-102

Scola, R. (1975) 'Food markets and shops in Manchester 1770-1870', *Journal of Historical Geography,* 1, 153-68

Taylor, A.J. (1975) *The standard of living in Britain in the industrial revolution,* Methuen, London

Thompson, E.P. (1971) 'The moral economy of the English crowd in the eighteenth century', *Past and Present,* 50, 76-136

Wild, M.T. & Shaw G. (1974) 'Locational behaviour of urban retailing during the nineteenth century: the example of Kingston upon Hull', *Transactions of the Institute of British Geographers,* 61, 101-18

Wild, M.T. & Shaw, G. (1975) 'Population distribution and retail provision: the case of the Halifax-Calder Valley area of West Yorkshire during the second half of the nineteenth century', *Journal of Historical Geography,* 1, 193-210

Willan, T.S. (1976) *The inland trade: studies in English internal trade in the sixteenth and seventeenth centuries,* Manchester University Press, Manchester

Wilson, C. (1965) 'Economy and society in late-Victorian Britain', *Economic History Review,* Second Series 18, 183-198

Yamey, B. (1954) 'The evolution of shopkeeping', *Lloyds Bank Review,* New Series 31, 31-44

Chapter 7

THE ROLE OF RETAILING IN THE URBAN ECONOMY

Gareth Shaw

Introduction

Despite a number of recent publications, the study
of retail structures and other associated activities
remains a relatively neglected issue when compared
with other facets of life in the nineteenth-century
city. In particular, knowledge about the internal
structure of commercial land use patterns, and the
role they played in shaping spatial aspects of the
urban economy, is at a rather basic level.

Research, principally by economic historians,
has begun to focus on the development of distribution systems, and more recently on the timing and
growth of the fixed-shop form of retailing. However, present opinion is divided between those
workers who suggest that the distribution system in
Britain prior to 1850 was of a primitive nature, and
that major changes only occurred after this date
(Jefferys, 1954), and others who believe that the
timing of change can be pushed back at least to the
latter part of the eighteenth century (Burnett,
1966; Alexander, 1970). This debate has not only
stimulated the study of retail change but has also
provided an important platform for the discussion of
other aspects of retail activity, notably food
supply and consumption patterns in nineteenth-
century towns (Blackman, 1963, 1967).

Recently, geographers have developed a further
research theme; the locational behaviour of shops
at the intra-urban level. Wild and Shaw (1974)
emphasized the links between changes in retail
structure and city growth, by tentatively describing
a three stage sequence of shop development. Subsequent research has further highlighted the various
processes affecting the locational decisions of
shopkeepers and especially the timing of retail
suburbanization in a number of settlements during

the nineteenth century (Wild and Shaw, 1979).
Whilst these studies have drawn attention to a
number of aspects of urban retail provision, they
have also raised additional issues for future
research. For example, there has been little
research on the possible relationships between
urban social change and corresponding shifts in
retail structure in nineteenth-century Britain; a
theme only really explored by Berry (1963) in the
context of the post-industrial North American city.
Berry's work on Chicago represents an extension of
existing urban ecological studies, since the processes of demographic and retail transition are seen
as synonymous. This link between urban ecology and
retail studies is important, as it gives the latter
an established framework which may aid the understanding of commercial land use patterns.

Urban growth and retail change

Changes in the structure, organization and location
of retail facilities are stimulated by changes in
population, per capita income and consumer mobility.
Often the combination of these demand variables, and
the corresponding reaction of the distribution
system, produces recognizable phases of retail
development. Bucklin (1972) presents a model of
such stages in which the lowest levels of economic
development are characterized by diffuse and small-
scale purchasing power, with the retail system being
dominated by periodic markets, which have the lowest
operating costs in such environments (Figure 7.1).
Increasing economic and urban growth leads to the
concentration of purchasing power, which favours
the establishment and growth of permanent markets
and craftsmen-retailers operating from fixed shops.
The latter group develop primarily because poor
levels of inter-urban transport limit areas of
supply, and favour the production of consumer goods
on a small and localized scale. Finally, as
transport improves and more sophisticated manufacturing technology allows the production of a greater
number and variety of products, the distribution
system becomes fragmented into distinct producers,
wholesalers and retailers, and large-scale retail
institutions begin to emerge.
     At the start of the nineteenth century many of
the initial changes suggested by this model of an
emerging retail system had been completed. Fairs
had ceased to be major centres of exchange, and
craft guilds had lost all power over the rapidly

172

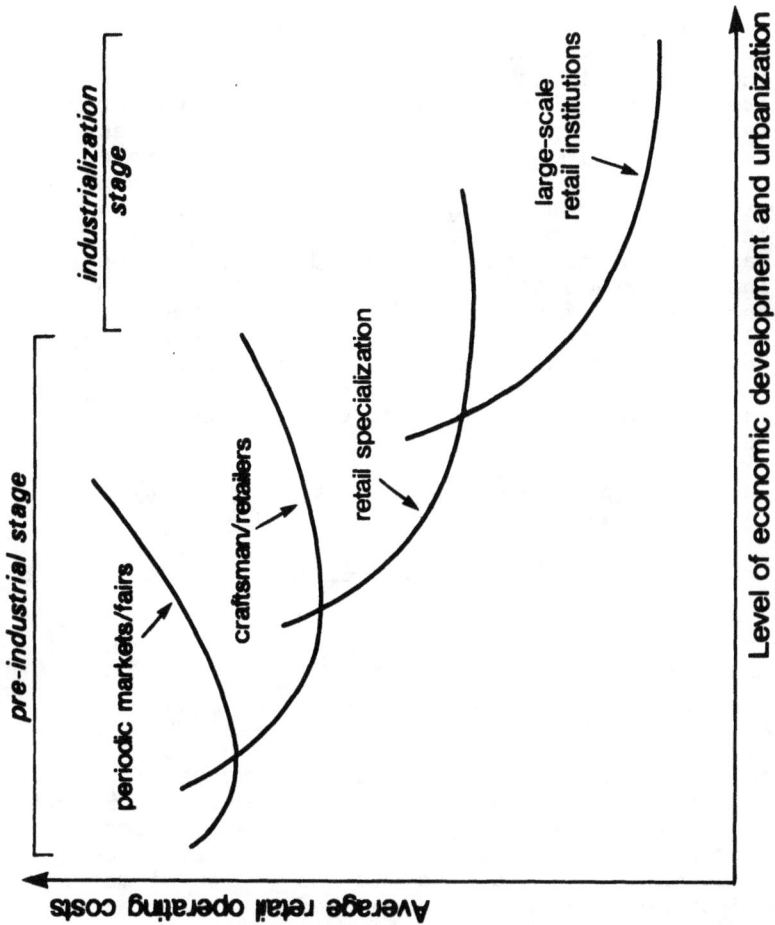

Figure 7.1 Stages of
retail development

173

emerging class of shopkeepers (Westerfield, 1951).
There seems little doubt from available evidence
that, by the first quarter of the nineteenth century,
shops had already become important supply outlets in
a wide range of settlements (Alexander, 1970; Wild
and Shaw, 1975). Indeed, in many towns the fastest
rates of shop increase occurred between the end of
the eighteenth century and 1820, although in most
areas shop numbers continued to rise considerably
faster than population until the mid-nineteenth
century (Table 7.1). Thus most early nineteenth-
century towns experienced an increase in shop provision, as measured by the average population per
shop. For the ten selected centres used in the
calculation of Table 7.1, population per shop fell
sharply from 136.3 in 1801 to 71.3 in 1821, but
then less dramatically to 56.7 in 1851 and only
slightly to 56.0 by 1881. We must not assume,
however, that such increases in shop provision were
achieved steadily, and at certain times shop growth
may not have always matched that of population
(Scola, 1975).

Table 7.1: Population and shop growth in selected urban
centres[1]

	Population growth (%)	Shop growth (%)
1801 - 1821	49.5	185.7
1821 - 1851	96.0	146.7
1851 - 1881	69.2	71.1

Concealed behind this information on shop
growth are other important developments in retail
change; in particular the changing roles of markets
and itinerant traders in the urban economy. In the
same way that shop development was stimulated by the
concentration of population and purchasing power in
towns and cities, so too was the growth of markets
and itinerant traders encouraged. Thus, markets
became increasingly important as retail centres
supplying the growing numbers of town dwellers with
food. This changed role led to a separation of
market functions, since the long established wholesale markets could not cope with large-scale retail
functions. This process is clearly illustrated in
London where, at the end of the nineteenth century,
the central markets were almost entirely wholesale
(Buzzacott, 1972).
    Urbanization also boosted the importance of
itinerant traders, whose numbers increased significantly after 1830 and continued to rise throughout

174

the rest of the century (Table 7.2).   The transient
nature of this occupation may well preclude accurate
enumeration, perhaps leading to significant under-
estimation of traders as suggested by Mayhew (1850),
but total numbers almost certainly fluctuated in
response to changes in the general level of employment, as unemployed workers from manufacturing and
agriculture moved into petty trading in hard times
(Lewis, 1954).   Even taking into account such
factors, it is still possible to assess the
importance of this form of retailing using census
data.   The growth in numbers recorded in Table 7.2
was primarily associated with urban development,
since it was the heavily urbanized counties that had
the largest numbers of traders (Figure 7.2).   The
rural fringes, particularly south-west England, had
by far the lowest proportions of itinerant traders,
a situation further intensified by an absolute
decline in the established country trade.   The
majority of rural hawkers and pedlers operated on a
larger scale than their urban counterparts, and unlike urban hawkers they took out a licence
(Alexander, 1970).   The number of licences issued
thus to some extent provides a measure of rural
trade, which reached a peak of 7,479 traders in 1830
but had declined to 5,762 by 1843 (British
Parliamentary Papers, 1800-1844).

Table 7.2   The number of itinerant traders in England and Wales

1831	1841	1851	1861	1871	1881	1891
9,459	14,662	25,747	37,735	49,775	47,111	58,939

Source: Census of England and Wales, 1831-1891

     The three main retail forms of shops, markets
and itinerant traders had all undergone some degree
of change by the first half of the nineteenth
century, each stimulated largely by urban growth.
However, the distribution system still remained
fairly basic in the sense that large numbers of
people were employed, mostly in very small units
(Shaw, 1976), and the distributive trades, like many
other aspects of the early-Victorian economy, were
characterized by the substitution of labour for the
relatively scarce resource of capital.   In mid-
Victorian Britain, continued urban growth and
expansion of the capitalist sector of the economy
gave renewed opportunities for the development of
large-scale retail organizations, although these
latent economic forces were not fully released until

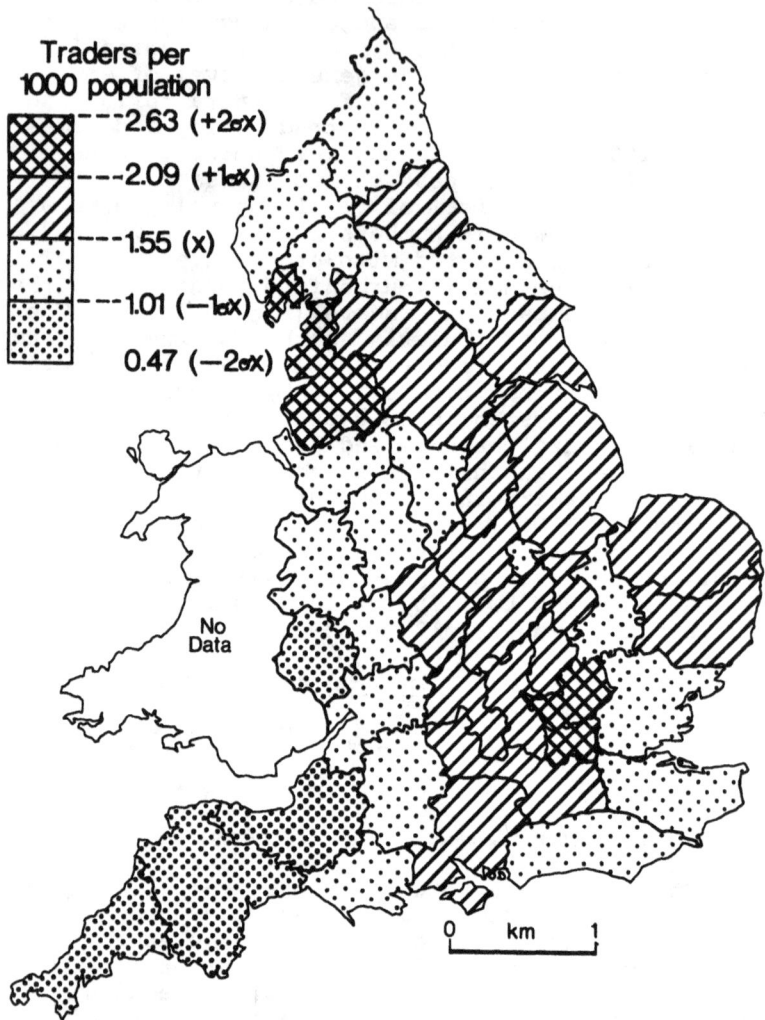

Traders per
1000 population

2.63 (+2σx)

2.09 (+1σx)

1.55 (x)

1.01 (−1σx)

0.47 (−2σx)

No Data

0   km   1

Figure 7.2   The distribution of itinerant traders
in England in 1861

improvements in inter-urban transportation and changes in product technology occurred.

These later changes in the distribution system took the form of a fragmentation of retail outlets and in particular the evolution of new forms of organization including co-operative societies, department stores and multiple retailers. The origins of the former seem uncertain, but in Britain they were established in limited numbers between 1850 and 1870 (Adburgham, 1964; Pasdermadjian, 1954; Harrison, 1975). Department stores were simply an attempt to gain economies of scale at the shop level by increasing the range of products sold from any one location, while multiple retailers also achieved organizational economies but through the development of large numbers of branch shops selling standardized products (Jeffreys, 1954; Shaw, 1978).

Identifying retail configurations

Researchers have been reluctant to identify hierarchies of shopping centres within the nineteenth-century and our knowledge of retail centres is mainly confined to simple representations of urban retail patterns such as those that distinguish between linear and nucleated components (Figure 7.3). This is partly because of a wish to avoid superimposing contemporary models of shopping centres onto quite different urban environments, but there has also been some mistrust of the classification methods used to identify and rank retail centres in twentieth century towns.

There are two main approaches to the classification of retail configurations. One uses the hierarchical concepts inherent in central place theory and differentiates on the basis of the size and functional status of shopping centres. A functional index is then calculated which is used to rank shopping centres. The second method is more broadly based and uses such criteria as the age of the centre, its form, functional role and size to discriminate between retail complexes. The application of these two methods and their relevance to our understanding of urban retail structures is, however, still far from clear despite considerable problems in their potential application to the description of commercial patterns in nineteenth-century cities.

Work on the locational behaviour of different types of shops in the nineteenth century does now

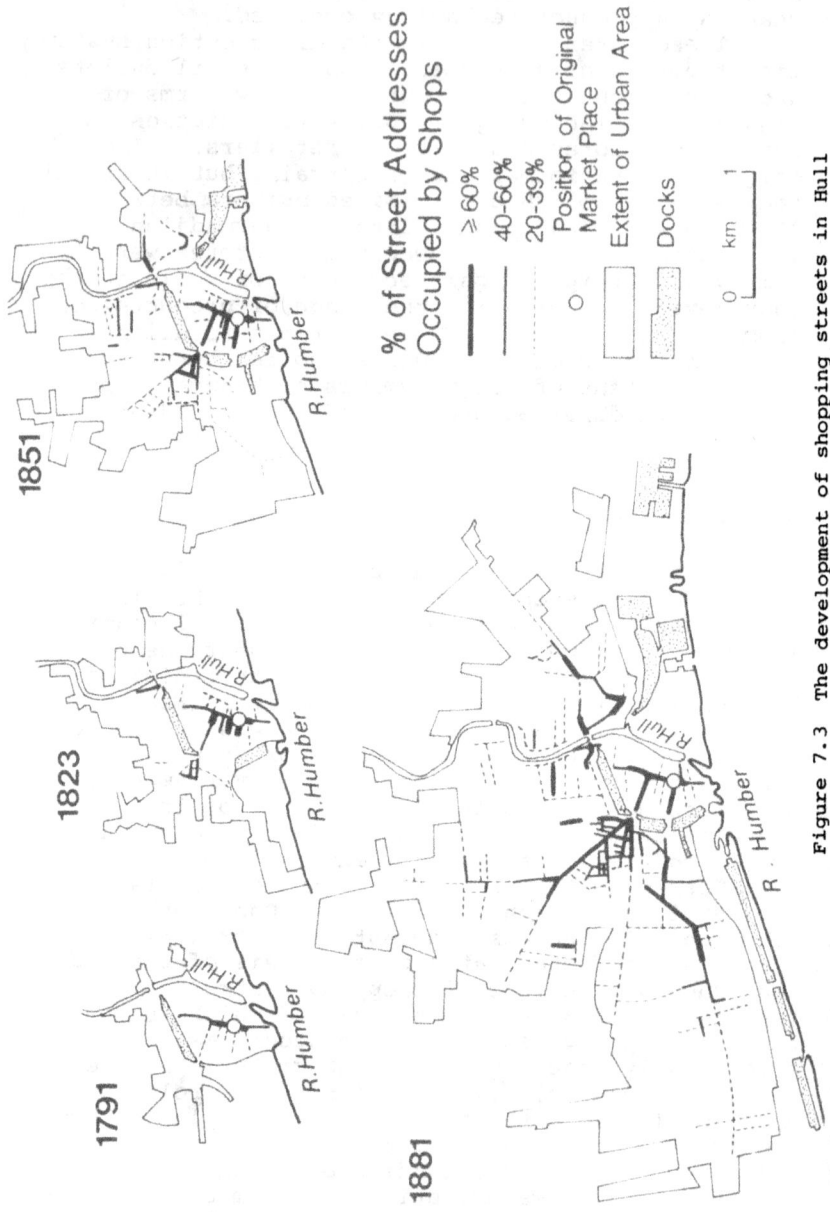

% of Street Addresses
Occupied by Shops

≥ 60%	
40-60%	
20-39%	
○	Position of Original Market Place
	Extent of Urban Area
	Docks

0   1
km

Figure 7.3 The development of shopping streets in Hull

provide some background knowledge to aid the
identification and classification of retail centres.
For example, shop types may be ranked according to
their locational behaviour (Table 7.3), and an
index of the functional structure of each shopping
centre may be constructed (Equation 1). Thus, low
order centres contain a proportionally greater
number of spatially dispersed trades, such as
'grocers and shopkeepers', and consequently have
low functional indices (Table 7.3A).   This
technique may be compared with an earlier method
suggested by Davies (1967), which is based on a
coefficient of location that depends on the number
of shops within the area of study (Equation 2 and
Table 7.3B).

Table 7.3: Alternative methods of ranking shop types
            (Blackburn, 1851)

IN ORDER OF:

(A) Locational behaviour[2]	(B) Davies' method
1. Ironmongers	1. Jewellers
2. Booksellers	2. Ironmongers
3. Jewellers	3. Booksellers
4. Pawnbrokers	4. Pawnbrokers
5. Confectioners	5. Confectioners
6. Chemists	6. Chemists
7. Milliners and dressmakers	7. Milliners and dressmakers
8. Drapers and tailors	8. Butchers
9. Butchers	9. Boot and shoemakers
10. Boot and shoemakers	10. Drapers and tailors
11. Grocers and shopkeepers	11. Grocers and shopkeepers

$$\text{Equation 1} \qquad S.I. = \frac{s \times 100}{T}$$

where	S.I.	=	structural index
	s	=	number of shops in centre
	T	=	total shop structure of centre

$$\text{Equation 2} \qquad c = \frac{t \times 100}{T}$$

where	c	=	location coefficient
	t	=	one outlet of function t
	T	=	total number of shop outlets of function t within study area

These two indices produce only slightly
different rankings of trades when applied to data
from nineteenth-century Blackburn (Table 7.3).
However, the former index is based on the sounder
principle of the locational behaviour of shops,
whilst the latter technique suffers from the fact

that the ranking of shop types, and hence the
classification of centres, is affected by the
initial number of shop types used. For example, if
a more generalized grouping of trades is used a
different 'locational coefficient' would emerge.
Furthermore, McEvoy (1968) has shown that such
variations significantly affect the final ranking
of shopping centres. This would appear to be a
major drawback to the use of such methods.
    More fundamental problems have also been
recognized, noteably the fact that such narrow-
based index techniques tell us nothing about the
other characteristics of shopping centres. Work
from both North America (Berry, 1963) and more
recently Britain (Davies, 1974), points to important
functional differences between nucleated and
linear shopping configurations in contemporary
cities. Such differences were probably not
apparent in the early- and mid-nineteenth century
because of low levels of consumer mobility and the
limited development of linear shopping complexes
(Figure 7.3). However, the arrival of tramways
stimulated ribbon developments, often with distinctive functional roles, and there is some
evidence to suggest that such areas provided sites
for the majority of early multiple-shop firms
moving into suburban areas (Dunlop, 1925; Shaw,
1978a). Before such detailed questions can be
fully answered, more information is needed on both
the ages of shopping centres and on their floor-
space characteristics.

The structure and distribution of nineteenth-century shopping centres

The above discussion suggests that more attention
needs to be given to the structure of retail
centres, especially the relative sizes of different
shops. However, calculation of usable shop
floorspace is extremely difficult, since only the
ground-floor area of shops can be calculated from
large-scale maps. Although information is sparse
it does seem that, after the mid-nineteenth century,
retailers were becoming increasingly conscious of
the importance of larger stores as many shopkeepers
attempted to achieve scale economies by increasing
the number and range of products on sale. One of
the earliest means of achieving such increases in
retail floorspace was through the amalgamation of
adjoining shop premises; a process that increased

180

in importance during the second half of the nineteenth century, particularly in city centres
(Table 7.4).

Table 7.4: The amalgamation of shop premises in selected
            towns[3]

Ratios of amalgamated shops:

	1823	1851	1881
In central areas	1:378	1:78	1:13
Over all town	1:600	1:156	1:32

Using the structural index (Equation 1) it is
only possible to identify retail centres on the
basis of numbers of shops and trade types. The
relationship between the size of centres, in terms
of shop numbers, and their functional importance
within the urban hierarchy is illustrated in
Figure 7.4.   This shows trends comparable to those
intra-urban hierarchies identified in a number of
studies of twentieth-century cities (Garrison,
Berry and Marble, 1959; Berry, 1963; Davies, 1974),
but interpretation of the relationships between
size and structure, and the identification of an
intra-urban hierarchy, poses the same problems as
in contemporary studies.   The most noteable
difficulty is that of classifying shopping centres
into various groups.   This is usually based on
distance measurements between points on the graph,
but doubt has been cast on this methodology,
largely due to inconsistency of application (Beaven,
1977).   However, such problems are less acute in
this exploratory work, and the classification of
centres is based on interpoint distances giving
three major groups of centres together with the
city centre shopping area (Figure 7.4).
The distribution of these shopping centres
highlights the role of the small local centre in
late-nineteenth century towns (Figure 7.4).   For
example, in Hull in 1880 there were 38 local
shopping centres scattered over a net built-up area
of 771 hectares, giving an average net density of
around one per 20 hectares, although the greatest
number of local centres was in the inner suburbs,
particularly the area urbanized before the mid-
nineteenth century (Figure 7.3).   In addition, more
localized needs were provided for by over 500 shops
located outside recognizable centres.   The five
'third order' shopping complexes were mainly
emerging ribbon developments, as can be seen by

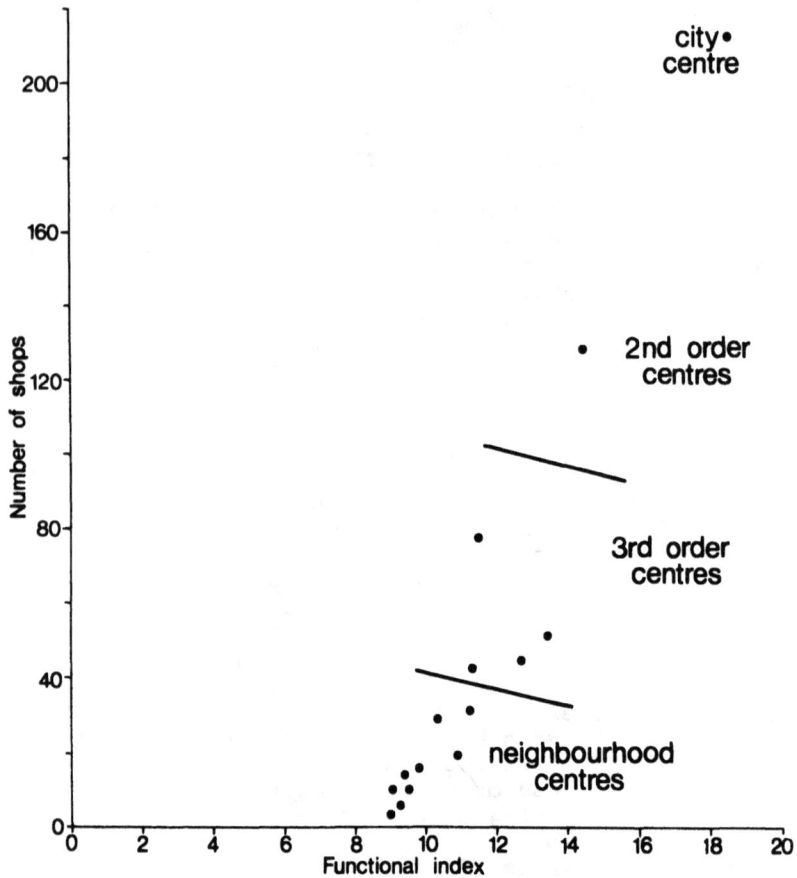

Figure 7.4   The relationship between shops
and retail structure in Hull

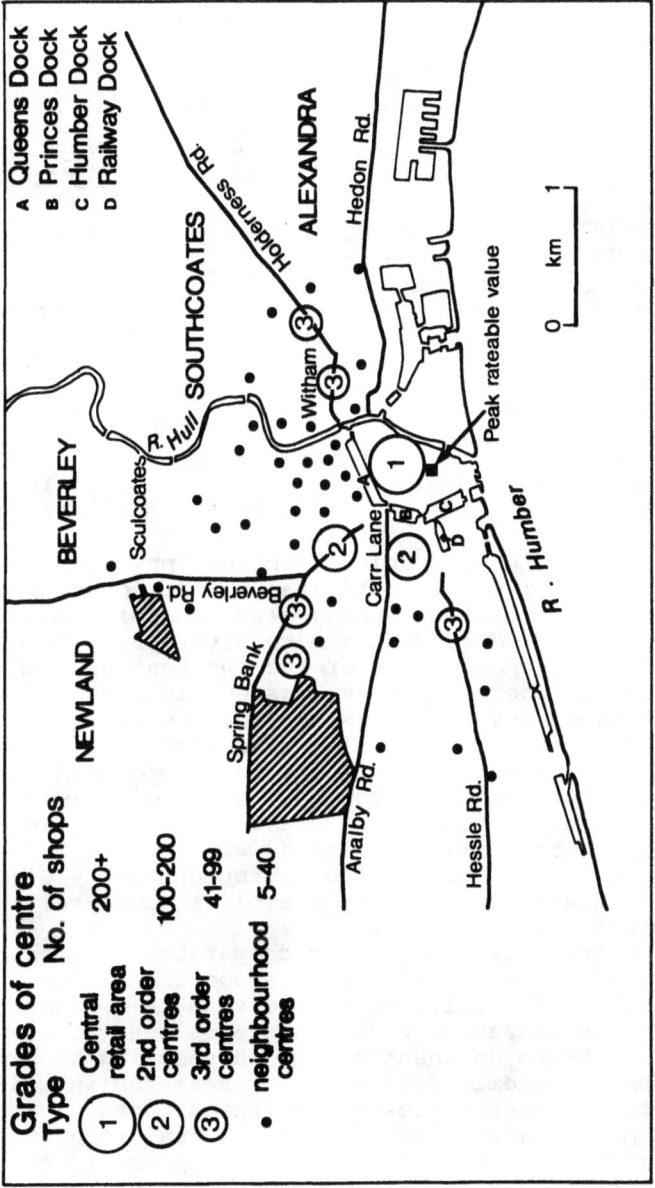

Figure 7.5 Shopping centres in Hull in 1881

comparing Figures 7.3 and 7.5, while the main
structural differences between the local and the
'third order' centres related to the degree of shop
provision.   Even the largest of the local centres
consisted mainly of low-order trades, although
noticeably not only food shops (Table 7.5) (Wild
and Shaw, 1974).

Table 7.5 The structure of some typical nineteenth-century
           shopping centres

| Trade types | Grades of centre | |
	Local	Third order
Ironmongers	0	1
Booksellers	0	1
Jewellers	0	0
Pawnbrokers	0	0
Confectioners	0	3
Chemists	0	4
Milliners/dressmakers	3	3
Drapers/tailors	0	8
Butchers	3	14
Boots and shoes	5	4
Grocers/shopkeepers	14	33
	25	71

The two 'second order' centres, physically
separated from the central area by the docks, both
evolved from smaller working-class shopping complexes as they developed a more diversified retail
structure.   The largest of the two centres, that
centred on Prospect Street, was eventually to
become an extension of the city centre shopping
area (Figures 7.3 and 7.5).   The growth of these
two fairly large retail centres close to the
central area, reflects the low levels of consumer
mobility among the working classes and the growth
of such centres may be compared with similar, if
somewhat smaller, developments recognized by Giggs
(1977) in a study of Barry's evolving commercial
structure.
Unfortunately, no other comparable studies
exist and there is still much scope for further
research.   Initially, this needs to be focused on
three main issues.   First, there is need to
examine the relationships between shopping centres
and socio-economic conditions.   Research in contemporary cities suggests that the structural composition of centres in low-status areas is quite
different from that of other social areas (Berry,
1963; Garner, 1966).   Low-status centres usually
contain only a limited variety of trades in

184

relation to the total number of shops, and some evidence for this relationship can be tentatively recognized in Figure 7.4 and Table 7.5. Furthermore, changes in the characteristics of shopping centres in response to social and demographic change also requires attention. At present, research has not progressed beyond the measurement of shop provision, which appears to be highly variable in the nineteenth century (Shaw, 1976), and the next stage must be to relate different aspects of shop provision to socio-economic conditions. A second area requiring additional work is the changing nature of the intra-urban shopping hierarchy, and the evolution of retail areas. It may even be possible to test the hypothetical model of shopping centre development suggested by Davies (1972), and if this is substantiated it may be feasible to develop a long-term model of shopping centre development comparable to that outlined by Vance (1962) for North American cities.

A final issue that requires more study is the role played by street markets and itinerant traders in nineteenth-century retailing. Although geographical studies have focused mainly on shops, it has already been shown that other elements of the distribution system were significant in nineteenth-century cities. Such factors undoubtedly complicate attempts to assess the relative importance of shopping centres. It is clear that, in the 1830s and 1840s, unregulated local street markets played an important role in many inner urban areas. However, due to data deficiencies, it is not possible to estimate the size of these features. Furthermore, the distribution of itinerant traders is almost impossible to assess because of the very nature of their trade. It would seem, however, that these different forms of retailing were not mutually exclusive in their location, and available evidence suggests that shopkeepers and street hawkers were to a certain extent dependent on each other in early working-class suburbs. Functional linkages can also be recognized between shops and street traders, as many of the small-scale costermongers and hawkers occasionally bought their products from larger shops (Alexander, 1970).

Retail development in central areas

The majority of existing research has been concerned with the development of suburban retailing,

while the study of commercial land use and shopping
provision in central areas has been relatively neglected.   The significance of central markets and
their changing status has received some attention
(Scola, 1975; Grady, 1976), but other facets of
city centre retailing have been almost completely
neglected, particularly the development of large
shops and department stores and their effects on
land-use patterns.   The impact of such developments
was experienced initially in the larger towns and
cities, which contained the high numbers of con-
sumers necessary for the growth of these retailers.

The emergence of London's West End as a
fashionable retail centre during the early-nineteenth century, provides a case study of both the
evolution of department stores and their effect on
commercial land use.   Three main phases can be
recognized in the development of department stores.
The earliest was the construction of bazaars
during the 1830s and 1840s, which were the
fashionable equivalents of market halls in the
sense that they were large buildings where individual retailers could rent a stall (Adburgham, 1964).
In mid-nineteenth century London about 10 such
bazaars operated, with the majority being found in
the expanding areas of Oxford Street and The Strand
(Figure 7.6).   The second phase of development was
represented by those retailers who had started to
increase their premises through the process of shop
amalgamation.   In the West End there were two
types of shop that grew in this way; the long
established firms that increased in size fairly
slowly, and those that emerged after 1850 and grew
more rapidly.   The former were generally located
in the heart of fashionable London, whilst the
latter tended to be more widely dispersed and many,
such as William Whiteley's 'Universal Provider' in
Westbourne Grove, were responsible for opening up
new streets as shopping areas.

The ultimate aim of many of the early department stores was to obtain an 'island' site or
street block by buying up adjoining buildings.
This would give both the floor space needed to
expand, and help eliminate competititon from nearby
shops.   This process was, however, difficult,
since physical expansion usually required the
removal of large numbers of smaller shops.   For
instance,before Bourne and Hollingsworth could gain
control of a street block for their Oxford Street
store, they had to buy up 30 premises for a ground
site of 900 square yards.   The desire for these
'island sites' was often so strong that some firms

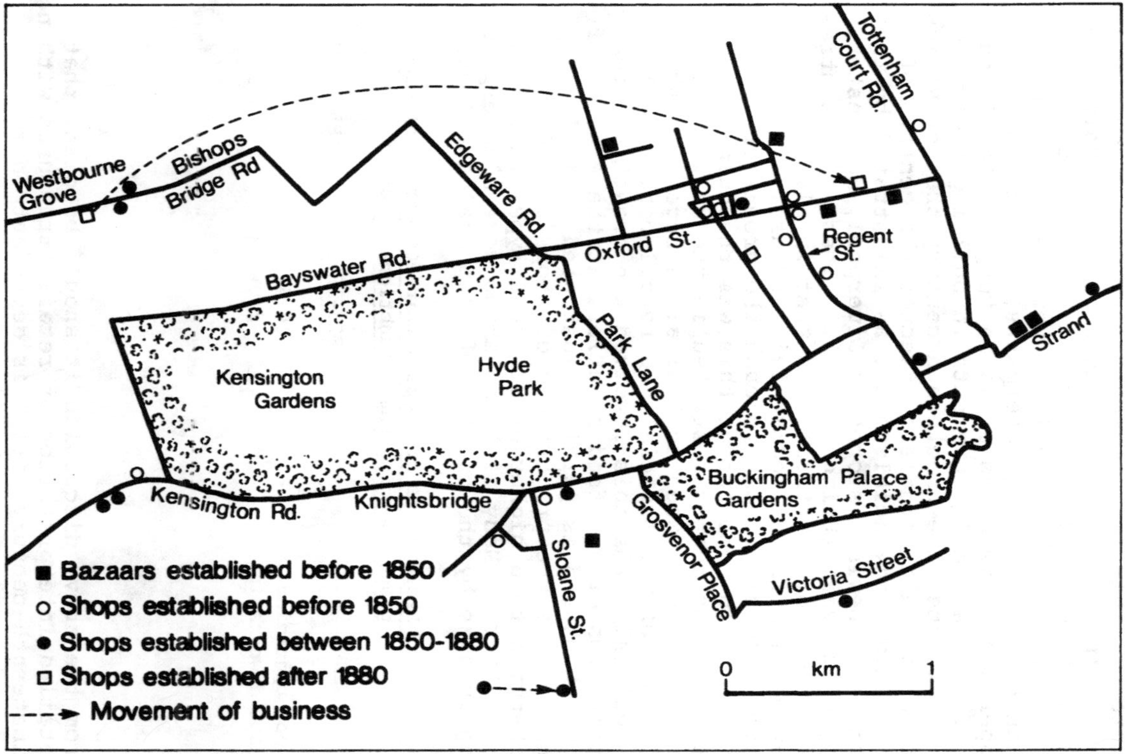
Tottenham Court Rd.
Westbourne Grove
Bishops Bridge Rd
Edgeware Rd.
Oxford St.
Regent St.
Bayswater Rd.
Strand
Park Lane
Kensington Gardens
Hyde Park
Buckingham Palace Gardens
Kensington Rd.
Knightsbridge
Grosvenor Place
Sloane St.
Victoria Street
0 km 1
Bazaars established before 1850
Shops established before 1850
Shops established between 1850-1880
Shops established after 1880
Movement of business

Figure 7.6 The development of department stores in London's West End

were persuaded to change locations; as in the case of Peter Jones who transferred his drapery business to a larger site in Sloane Square (Figure 7.6).

The third phase of growth occurred after 1880 with the construction of purpose-built department stores. This trend reached its peak in the West End during the early-twentieth century, typified by the opening of Selfridges in 1909. These later stores laid more emphasis on vertical development than their earlier counterparts, and made use of upper floors as sales areas in response to rising land values and improvements in building technology.

In the late-nineteenth century the activities of large-scale retailers affected the centres of most large provincial towns, as illustrated by the growth of Lewis's stores in Liverpool, Manchester and Birmingham (Briggs, 1956). These city centres were undergoing a transition that involved the rationalization of commercial land use, as small shops were swept away through the growth of large stores. To some extent this was the culmination of a process that had started in the early-nineteenth century, when central areas began to take on the role of more specialized retail centres. In functional terms this involved the growth and concentration of non-food retailers at the expense of food shops and is partly illustrated in Table 7.6, which shows the proportions of shops upgrading (changing to a higher order trade type) in 10 towns and cities. Many of these shops were also expanding in size by the process of shop amalgamations.

Table 7.6: Proportions of shops within 200 metres of market place changing their functional status[4]

	1823-1851	1851-1881
Shops upgrading	16.4%	39.7%
Shops downgrading	2.0%	4.2%

From the above discussion it should be clear that detailed interpretation of retail structures within nineteenth-century cities is dependent on an increased level of empirical analysis. To a large extent such research is hampered by the nature of existing data sources. For example, any work seeking to explain the growth and diffusion of department stores would need to slowly piece together the surviving, but as yet largely unused, records of many provincial retail firms. A lead has already been given by economic historians who have examined some individual business histories (Briggs, 1956; Lambert, 1938) but such work now needs to be placed on a more systematic basis and given a spatial dimension by historical geographers. Furthermore, where data exist in relative abundance, as in the case of trade directories and the occupation abstracts of the census, their reliability remains largely untested (Norton, 1950; Shaw, 1978b).

The reliability of directories for the reconstruction of past retail patterns needs to be examined from two viewpoints. First, their overall level of accuracy must be assessed and, secondly, the extent to which they allow information to be accurately located must be determined. Checking the accuracy of directory information is not easy since they cannot meaningfully be compared with other data sources such as rate books, because they exist in completely different formats. One particular problem is the fact that many shops, especially in working-class areas, were not rated as retail establishments since business was conducted from the front room of a house and consequently these were listed as purely residential properties in the rate books.

Such anomalies may be illustrated by examples from Beverley, an established market town, and Hull, a rapidly growing industrial and commercial centre. In Beverley the number of shops given in *White's Directory* of 1887 was 5.6 per cent below the number enumerated in the town rate books. In contrast, in Hull the same directory contained 5.2 per cent more shops than those in the rate books. This suggests that in Hull, where there were a greater number of working-class shopping centres and small 'front parlour' shops, directories provide a more accurate measure. However, at street level this difference between shop numbers is complicated and more variable, as illustrated in Figure 7.7 for a ten per cent sample of shopping streets in Hull. In addition, rate books do not usually give information on the type of shop, and hence comparisons with directories can only be made on the total number of shops rather than on a trade by trade basis.

Comparisons have also been made between directories and census data, using both occupation abstracts and enumerators' books (Wild and Shaw, 1979). Although census occupation abstracts only become useful after 1841 when a distinction is made

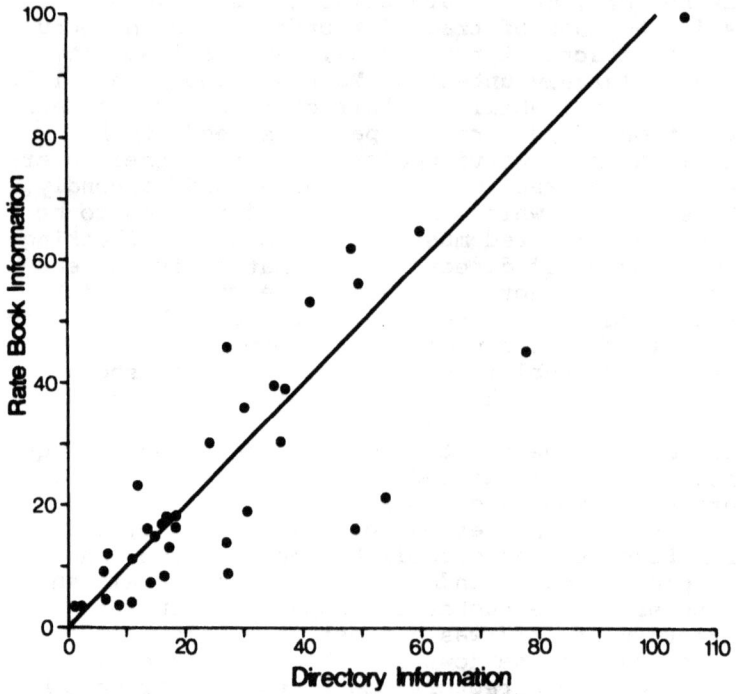

Figure 7.7   Numbers of shops enumerated by White's
Hull directory and town rate books by streets

between employers and employees, these sources do
have the potential advantage of allowing comparisons
to be made at an individual trade level.
Difficulties will still arise however because of
the different trade classifications used and the
diverse areas over which data was collected.

The reliability of directories can be assessed
in at least one further way; by checking the background and type of compiler/publisher involved in
producing the directory. For example, many of the
regional and national firms employed full-time professional agents responsible for systematically
collecting and recording information. One of the
earliest and most reliable of these firms was that
run by Edward Baines who published directories for
Lancashire and Yorkshire between 1822 and 1825. In
contrast to these nationally produced directories
were those issued by local printers and publishers,
which were usually small concerns often producing
only one directory. For example, of the 81
directory publishers operating in Lancashire between
1790 and 1900 almost 52 per cent produced only one
directory. The great majority of these directories
were produced by non-specialists and, as a result,
were less carefully compiled.

A second aspect of directory reliability, the
ability to accurately locate information, is rather
easier to assess. The consistency with which
street addresses were given often depended on the
reliability of the compiler/publisher, the date of
publication and the size of the settlement the
directory covered. However, in most directories
the address coverage of shops for any particular
town was usually incomplete (Table 7.7). In
addition, the omission of street addresses for
shops was often geographically biased towards the
growing suburbs, places where the very rapid nature
of change made accurate data collection difficult
even for experienced directory agents. Furthermore,
there appears to have been some variation in
address omissions by shop type (Table 7.7). Both
these factors may help account for the variations
in shop counts illustrated in Figure 7.7.

Table 7.7: Address omissions in White's Hull directory (1881)

| Trade | Shops without addresses | | Percentage falling |
	Total	Percentage	in suburbs
Grocers	50	19	62
Butchers	18	11	56
Footwear	8	3	0

This chapter has attempted to review existing research and, more importantly, to highlight some potential areas for further study.   Recent progress has been focused particularly on explaining the development of shops, the role of markets and the suburbanization of shopping facilities.   The majority of these studies have drawn attention to the difficulties involved in using trade directories to reconstruct past retail environments, and this essay has revealed further problems which need to be considered in future work.   However, despite their limitations it seems that trade directories will remain the main source of data for reconstructing nineteenth-century retail patterns and further research on their general levels of accuracy should make them much more acceptable data sources for historical geographers interested in the retail structure of nineteenth-century towns.

Notes

1.   The towns used were Hull, Halifax, Leeds, Huddersfield, Oldham, York, Rochdale, Blackburn, Preston and Lancaster.
2.   These figures are based on the locational index developed by Wild and Shaw (1974) which is itself based on the Lorenz curve values of shop distribution.
3.   The towns used were Halifax, Huddersfield, Hull, Oldham, Rochdale and York.
4.   Status refers to the ranking of trade types as in Table 7.4, and covers the towns listed in note 1.

References

Adburgham, A. (1964) *Shops and shopping, 1800-1914,* Allen and Unwin, London
Alexander, D. (1970) *Retailing in England during the industrial revolution,* Athlone Press, London
Beavon, K.S.O. (1977) *Central place theory: a reinterpretation,* Longman, New York
Berry, B.J.L. (1963) *Commercial structure and commercial blight,* University of Chicago Research Paper No. 85, Chicago
Blackman, J. (1963) 'The food supply of an industrial town', *Business History,* 5, 83-97
Briggs, A. (1956) *Friends of the people,* Batsford, London
Bucklin, L.P. (1972) *Competition and evolution in the distributive trades,* Prentice-Hall, New York
Burnett, J.B. (1966) *Plenty and Want,* Nelson, London
Buzzacott, K.L. (1972) London's Markets: their growth, characteristics and functions. Unpublished Ph.D. thesis,

192

University of London

Davies, R.L. (1972) 'Structural models of retail distribution', *Transactions of the Institute of British Geographers*, 57, 59-82

Davies, R.L. (1974) 'Nucleated and ribbon components of the urban retail system in Britain', *Town Planning Review*, 45, 91-111

Davies, W.K.D. (1967) 'Centrality and the central place hierarchy', *Urban Studies*, 4, 61-79

Dunlop, W.R. (1925) 'A contribution to the study of London's retail meat trade', *The Economic Journal*, 35, 416-28

Garner, B.J. (1966) *The internal structure of retail nucleations*, Northwestern University Studies in Geography No. 12, Evanston

Garrison, W.L., Berry, B.J.L. and Marble, D.F. (1959) *Studies of highway development and geographic change*, University of Washington Press, Seattle

Giggs, J.A. (1977) 'The changing commercial structure of Barry, 1884-1976', Unpublished paper presented to I.B.G. Urban Study Group, September, 1977

Grady, K. (1976) 'The provision of markets in Leeds, 1822-1829.' *Publications of the Thoresby Society*, 54, 186-198

Harrison, M. (1975) *People and shopping*, Benn, London

Jefferys, J.B. (1954) *Retail trading in Britain, 1850-1950*, Cambridge University Press, Cambridge

Lambert, R.S. (1938) *The Universal Provider: a study of W. Whiteley*, Harrap, London

Lewis, W.A. (1954) 'Economic development with unlimited supplies of labour', *The Manchester School of Economic and Social Studies*, 22, 139-191

Mayhew, H. (1861-62, 1967 edition) *London labour and the London poor*, Cass, London

McEvoy, D. (1968) 'Alternative methods of ranking shopping centres', *Tijdschrift voor Economische en Sociale Geographie*, 59, 211-17

Norton, J. (1950) *A guide to the national and provincial directories of England and Wales (excluding London) published before 1856*, Royal Historical Society, London

Pasdermadjian, H. (1954) *The department store*, Newman Books, London

Scola, R. (1975) Food markets and shops in Manchester, 1770-1870, *Journal of Historical Geography*, 1, 153-68

Shaw, G. (1976) The geography of changes in retail trading patterns. Unpublished Ph.D. thesis, University of Hull

Shaw, G. (1978a) *Processes and patterns in the geography of retail change*, University of Hull Occasional Papers in Geography No. 24, Hull

Shaw, G. (1978b) 'The content and reliability of nineteenth century trade directories', *Local Historian*, 13, 205-9

Vance, J.E. (1962) 'Emerging patterns of commercial structure in American cities', in K. Norborg (ed.) 'Proceedings of the IGU Symposium in Urban Geography', *Lund Studies in*

Geography, 24B, 485-518

Westerfield, R.B. (1915) *Middlemen in English Business between 1660 and 1760*, Yale University Press, New Haven, Connecticut

Wild, M.T. and Shaw, G. (1974) 'Locational behaviour of urban retailing during the nineteenth century: the example of Kingston upon Hull', *Transactions of the Institute of British Geographers*, 61, 101-18

Wild, M.T. and Shaw, G. (1979) 'Trends in urban retailing: the British experience during the nineteenth century', *Tijdschrift voor Economische en Sociale Geographie*, 70, 35-44

PART FOUR

THE SOCIAL STRUCTURE OF NINETEENTH-CENTURY TOWNS

Although there remain many points of disagreement, differences in emphasis and in interpretation, perhaps the most striking feature of the four essays in this section is the degree of agreement which they eventually reach. Although approached from the points of view of historian, sociologist and geographer, and indeed aimed at rather different themes, an element of consensus seems to emerge about the most important topics to be tackled, the need for carefully-considered theoretical frameworks and the desirability of extending studies of residential differentiation beyond the limits imposed by the few years for which census enumerators' books are available. For instance, one topic explicitly considered in the first three essays in this section concerns the question of the modernity of the Victorian city. Richard Dennis is particularly concerned with the assessment of the degree to which there was significant change in the residential structure of nineteenth-century towns, suggesting that we may be chasing a shadow; Colin Pooley also questions the usefulness of the concept of a continuum of urban residential change, raising some of the theoretical and methodological problems which bedevil detailed comparisons; while David Cannadine focuses his attention on the potential (and confusing) implications of the modernity debate for the study of social relations in towns. Although there may remain small areas of overlap between the essays on this topic, such sections have not been removed as this would have caused imbalance within individual contributions, while the points of contact between authors are seen as essentially complimentary rather than unnecessarily repetitive.

In the first of the four chapters in this section Colin Pooley uses a relatively broad brush

in an attempt to outline the processes of change
which were operating within urban areas and the
theoretical frameworks most appropriate for their
study. Using a framework based on the concept of
choice and constraint in the residential location
decision, he presents a case-study of mid-Victorian
Liverpool. As with some of the other geographical
contributions to this volume, much of the theory
and methodology discussed is derived directly from
contemporary urban geography. Although this may
be useful, caution must be exercised when applying
contemporary concepts to past societies.

David Cannadine focuses his contribution on
one particular question arising out of the
residential segregation found in nineteenth-century
towns: the extent to which spatial separation
affected social relations and *vice versa*.
Cannadine's paper emphasizes the necessarily interdisciplinary nature of research into the social
structure of nineteenth-century towns, and highlights the extent to which apparently objective data
can produce very subjective and diverse interpretations. Given such disparity of interpretation, it is especially important that the questions
which are asked are soundly grounded in theory.
Some of the possibilities and problems of developing such theory are amply discussed in Cannadine's
paper.

Richard Dennis's paper concentrates on problems
in the study of change over time, changes both in
the structure of the city and in our levels of
analysis, ranging from the identification of
aggregate ecological social areas to the study of
individual behaviour patterns. He carefully
evaluates problems and possibilities for the spatial
study of different levels of community in nineteenth-century towns, expressing dissatisfaction
with much that has gone before and suggesting
directions in which future research might progress.

Lastly, Michael Anderson examines just one of
the measures of community networks that Dennis
mentions: the measurement of residential mobility
in towns. In this context he poses a number of
searching questions. Is the mobility process one
that was meaningful in the context of nineteenth-
century society? Are we asking the right questions
and collecting the best data? Do we know what
mobility really meant in both individual and
ecological terms in the Victorian city? Although
some of Anderson's questions may be unanswerable
given data constraints at the individual level, the
paper has implications extending outside the study

of residential mobility. For any historical
research to be meaningful the process being studied
must be shown to have been important in that
particular society and at that time in the past.
Its implications for different sections of that
past society must also be understood, and it must
be measured in such a way that its more significant
effects will emerge. As such remarks may be
equally applied to most of the topics that have been
discussed in this book, Anderson's 'sceptical
comments' provide a fittingly cautionary note on
which to conclude.

Chapter 8

CHOICE AND CONSTRAINT IN THE NINETEENTH-CENTURY CITY: A BASIS FOR RESIDENTIAL DIFFERENTIATION

Colin G.Pooley

Introduction

An awareness of residential differentiation in nineteenth-century British cities is not new.   Contemporary observers were concerned with the rapid
growth of towns in the nineteenth century (Vaughan,
1843; Weber, 1899) and with the way in which
different groups within society were becoming
increasingly segregated into particular areas of the
city (Engels, 1845; Booth, 1889; Flinn, 1965; P.P.
1845, XVIII).   These and other contemporary views
of Victorian urban structure have been reviewed in a
number of more recent works (Dyos, 1967; Cannadine,
1976).
     It was, however, the late-nineteenth century
American city that stimulated the first formalized
descriptions of urban structure.   Chicago, rebuilt
after a fire in 1871, rapidly developed a well-
defined core with peripheral suburban growth (Vance,
1976).   It was in this environment that Burgess
(1924) constructed his model of the development of
urban social structure in concentric rings, which
was to prove the stimulus for a continuing debate on
urban form and the processes causing particular city
structures to emerge (Hoyt, 1939; Harris and Ullman,
1945; Rees, 1970; Murdie, 1976).   The last thirty
years have seen a massive upsurge of research into
the development of urban social areas ranging from
the early work on Los Angeles (Shevky and Williams,
1949; Shevky and Bell, 1955) to complex factorial
ecologies of cities in most parts of the world
(Sweetser, 1962; Berry and Rees, 1969; Murdie, 1969;
Robson, 1969; Davies and Lewis 1973).   This.
literature has also been extensively reviewed elsewhere, including critiques of both the techniques
used (Dogan and Rokkan, 1969; Berry, 1971; Hunter,
1972; Berry and Smith, 1972; Mather and Openshaw,

1974; Clark, Davies and Johnston, 1974; Johnston, 1976) and the results obtained (Johnston, 1971; Timms, 1971; Johnston, 1972; Robson, 1973a).

Similar techniques have more recently been applied to the study of nineteenth-century urban structure (Goheen, 1970; Tansey, 1973; Lawton and Pooley, 1975b; Shaw, 1977). Given the wide range of variables which may be computed from the census enumerators' books of England and Wales for the five census years 1841-1881 (Wrigley, 1972; Lawton, 1978), it is possible to examine in detail the spatial structure of mid-Victorian towns and reassess the accuracy of contemporary observations on residential segregation in the nineteenth century. This essay seeks to advance discussion on residential differentiation in the Victorian city in three ways. First, it briefly assesses the levels of segregation that have been found to exist. Second, it examines the main processes that operated to cause residential segregation. Third, it explores the causes of residential differentiation in mid-Victorian Liverpool within a framework of choice and constraint in residential location derived from a review of current urban geographical theory.

Residential differentiation in the Victorian city

Recent analyses of nineteenth-century urban structure have tended to adopt one of two distinct approaches. On the one hand are those writers who stress the transitional nature of the nineteenth-century city, viewing it as part of a continuum of change from a pre-industrial to a modern post-industrial urban form (Goheen, 1970; Timms, 1971; Warnes, 1973). Such an approach tends to stress the differences between the nineteenth-century and the modern city, suggesting that twentieth-century models of urban development are inapplicable to the study of Victorian towns. Perhaps this view is most strongly stated by Ward (1975, 1976, 1980) who argues that, despite segregation at the extremes of society, the Victorian city in Britain and America was characterized more by a subtle intermixing of groups and classes than by the formation of distinctive residential areas. On the other hand it has also been suggested that, in Britain at least, many of the larger industrial towns had developed clear patterns of residential differentiation by the mid-nineteenth century. Their structure was such that the similarities to modern towns were greater than the differences and thus twentieth-century theories of urban development are likely to be applicable to the study of these Victorian towns (Lawton and Pooley, 1976; Cannadine, 1977).

The implications of these two approaches are discussed in two other contributions to this volume (Cannadine, Dennis), but before any assessment of nineteenth-century urban structure can be made a number of points must be clarified.   If the Victorian city is to be viewed as transitional between a 'pre-industrial' and a 'modern' urban form, then the structure of the pre-industrial and modern cities with which it is being compared must be clearly defined.   This, however, is rarely achieved. Sjoberg's (1960) model is usually taken as representative of the typical pre-industrial city, with the élite dominating the urban core and the poor dispersed to the periphery.   A modern city can be idealized as Murdie's (1969) model, where discrete dimensions of social status, family status and migrant or ethnic status are manifested in different spatial patterns which overlap to form a clearly differentiated urban structure, with the poor clustering around the edge of the Central Business District and the majority of the urban population being variously distributed within the suburban ring. Both these models are, however, over-simplifications of a complex reality.

While many pre-industrial cities were residentially differentiated they did not always conform to a simple centre-periphery model.   More frequently, particular clusters of occupational groups were apparent in the developing urban structure (Langton, 1975; Langton and Laxton, 1978) though these groupings did not necessarily reflect social-status segregation.   Many areas were socially mixed, with vertical segregation between floors of a single building the dominant influence (Vance, 1971); high-status residences could be found in many different parts of a town (Keene, 1979), while migrant groupings further complicated the urban structure (Clark, 1972).   In North America, towns with a pre-industrial economy also began to show clear occupational groupings with the richer merchants, least tied by constraints of transport, moving out beyond the urban fringe (Savanger, 1978).

Similarly, the structure of a twentieth-century industrial city cannot be simply defined.   Many different physical and economic circumstances can distort the classical models (Berry and Rees, 1969; Johnston, 1971), while the perceived character of an area can cause it to retain high status despite a position close to the low-status core (Firey,

1945). Variations in policies on housing provision can further complicate modern urban form (Vance, 1976), with public-sector housing in Britain proving a particular distortion (Robson, 1969). Furthermore, it has recently been argued that the most important forces controlling urban structure today are financial and institutional (Harvey, 1973, 1974, 1975b). These are rather different from the ecological forces deemed to produce the classical model of urban spatial structure with which nineteenth-century cities are usually compared. It is not surprising that the urban structure of both pre-industrial and modern cities is complex, but if the beginning and end of the hypothesized continuum of urban change cannot be clearly defined (Timms, 1971; Shaw, 1977), the position of the nineteenth-century city must remain unclear.

It should also be noted that city size and the scale of urban development are important considerations when comparisons are made between nineteenth-century towns. The decades of most rapid population growth occurred much earlier in Britain than in America (Weber, 1899; Law, 1967; Robson, 1973b), perhaps accounting for some of the discrepancies noted (Ward, 1975; Cannadine, 1976), but even in Britain rates of urban development varied markedly. While towns such as York and Lincoln largely escaped the direct impact of the industrial revolution (Armstrong, 1974; Hill, 1974), many others, and especially those such as Manchester, Liverpool and Leeds which inspired contemporary comments on their internal differentiation, were large industrial centres with populations of well over 100,000 by 1840 (Robson, 1973b). It seems likely that those towns which experienced rapid growth and industrialization in the early decades of the nineteenth-century were the first to develop residentially segregated social areas. Other towns may have retained a greater degree of intermixing between social classes and migrant groups until the later nineteenth century.

The development of an effective suburban transport network has been seen as a necessary pre-condition for large-scale residential differentiation (Ward, 1964, 1975; Warner, 1962; Warnes, 1970). That most towns did not gain a relatively cheap and efficient tram and suburban rail service until the late-nineteenth century (Kellett, 1969; Dyos and Aldcroft, 1969; Perkin, 1970) suggests a lack of residential differentiation in the mid-Victorian town. But even the large provincial city, though attaining a population of 500,000 or more, remained

202

relatively compact. Manchester could be comfortably traversed on foot in under one hour in the 1870s and for most smaller towns public transport was irrelevant. Many working men in regular employment would think nothing of a two-mile walk to work (Lawton and Pooley, 1975a), while the location of industry meant that much working-class housing could develop close to dock or canal-side factories. For some higher-status residents the flight to distant semi-rural suburbs had, in any case, begun long before the advent of cheap mass transport with horse-drawn omnibuses and private carriages providing the necessary transport. Although cheap suburban transport certainly allowed large-scale suburban sprawl in the late-nineteenth and early-twentieth centuries in both Britain and America, the lack of such transport did not prevent residential differentiation occurring in the physically smaller mid-Victorian city.

In comparing studies and assessing the validity of alternative theories, the constraints imposed by different techniques of analysis must also be considered. Most important are variations in the scale of spatial analysis and in the choice of variables. While analysis of enumeration districts may adequately reveal spatial variations in a large well-developed city, such a scale of analysis may obscure the smaller scale zoning found in a medium-sized town. In such towns analysis must be undertaken at the level of the street, the block or even the individual house, and it is essential that the most appropriate scale of analysis is chosen if residential patterns are not to be obscured (Pooley, 1973). Ecological measures of residential segregation, including the much-used indices of dissimilarity, can also produce spurious results if residential groupings fail to coincide with sub-area boundaries or if the number of individuals and sub-areas is small (Taeuber and Taeuber, 1965; Woods, 1976). The choice of variables used in most nineteenth-century urban studies is limited by the availability of census, map and rate-book evidence, but variables such as socio-economic status are open to widely differing interpretations which can severely affect comparisons. Although it is not appropriate to detail problems of census coding, classification and analysis in this essay, it is clear that the conventions adopted can have a considerable bearing on the results obtained (Wrigley, 1972; Lawton and Pooley, 1976).

While variations in definition and classification may account for some of the differing interpretations, empirical evidence does appear to be genuinely equivocal. For instance, Shaw (1977) couched his analysis of Wolverhampton within the framework of an hypothesized continuum of change during the nineteenth century, but concluded that in the mid-Victorian period this city exhibited many characteristics which were essentially modern; while even the small town of Chorley, Lancs, although characterized as having a 'transitional' urban form, contained distinctive occupational groupings from the early-nineteenth century, and the results of a factor analysis of social and economic variables for 1851 were not dissimilar to those from modern towns (Warnes, 1973).

For a large industrial city such as Liverpool it is easy to demonstrate that, by the mid-Victorian period, particular migrant, occupational and social-status groups had become segregated into different portions of the city and that distinct residential areas had emerged (Lawton and Pooley, 1976; Pooley, 1977). At the micro-scale, however, it cannot be argued that all of these areas were totally homo-geneous in character. Certain parts of the city - particularly the older central areas where craft industries and small workshops still remained - retained distinctly mixed social characteristics in 1871. Depending on the preferred theory and scale of analysis, Liverpool could equally validly be interpreted as a modern industrial city with a well-differentiated residential structure, or as a Victorian city still undergoing the transition from a pre-industrial to a modern urban form. From a different perspective Ward (1976), though arguing for an undifferentiated city, admits that at the street and block level distinctive clusterings of social, occupational and migrant groups did occur in even the smallest towns. Although in the classical ecological sense such groupings do not form social areas, they do indicate that the processes causing segregation and residential differentiation were operating, but at a lower level than in the larger industrialized city.

Perhaps differences between the two approaches are not as great as at first seems the case. Both can be supported by empirical evidence, but the interpretation of this evidence depends as much on the scales of analysis and the theoretical premises of the researcher as on any real differences between the towns studied. Most researchers agree that during the nineteenth century an increasing degree of residential differentiation occurred, linked with the growth of industrialization and urbanization.

The precise timing and nature of this differentiation varied from town to town: few residential areas were totally homogeneous, yet within most cities micro-scale groupings at a street or block level were apparent. Further cross-sectional case studies will enable the timing of change in cities of different sizes to be more fully documented, but perhaps it is time for urban historical research to move away from this somewhat sterile and static approach, which has tended to emphasize a false dichotomy within the development of urban structure. Instead, attention should be focused on the processes of change which manifested themselves sooner or later in the residentially-differentiated town described in successive models of twentieth-century urban development.

Processes of residential differentiation in the nineteenth century

For research into nineteenth-century urban residential areas to progress beyond a discussion of residential patterns and urban form at successive cross-sections of time, a number of processes require more detailed investigation.

First, the physical growth of the city and the evolution of housing areas must be investigated. As discussed elsewhere in this volume, house-building for the lower end of the market was severely limited in the mid-Victorian city; at best, only a skilled working man in regular employment could afford new housing (Bowley and Burnett-Hurst, 1915; Parry-Lewis, 1965; Gauldie, 1974). Thus, while middle-income bye-law terrace housing expanded, the stock of purpose-built lower-rent accommodation remained almost static in most towns after about 1850. Demand for cheap housing remained high, however, especially following the Irish famine migrations (Chapman, 1971; Gauldie, 1974; Burnett, 1978). This demand could only be met by increased multiple-occupancy or through the low-rent sector expanding at the expense of other housing. Only where existing middle-status housing could filter down to those at the bottom of the housing market did the physical extent of housing for the poorest classes expand. The development of new middle- and high-status housing around the urban periphery allowed inner-city property to be vacated by its former residents who moved to the suburbs, while economically less well-off families took their place.

Relatively little is known about the detailed operation of the housing market and the filtering process in the Victorian city, or the extent to which distinct housing classes and sub-classes become associated with spatially-discrete areas. While the exact classification used by Rex and Moore (1967) is not appropriate to the nineteenth century, the concept of housing classes, each of which is open only to certain sectors of the population and has a clear spatial expression, is one which needs much more careful investigation. Greater understanding of the nineteenth-century housing market, including an assessment of the effectiveness of the filtering process in providing increased quantities of low-rent accommodation, is clearly fundamental to future studies of the processes underlying the formation of residential areas.

Secondly, the development of residential differentiation must be seen within the context of individual locational decision-making in a rapidly-expanding urban population. As already suggested, urban populations tended to expand more rapidly than the available housing stock, leading to massive over-crowding and high levels of multiple-occupancy in most towns (Gauldie, 1974). The urban newcomer would thus be faced with a complex decision on residential location. Furthermore, the in-migrant to urban areas competed with existing residents for the limited housing stock: few urban dwellers remained in the same house for long in the nineteenth century, and both migrants and established residents must have been continually re-evaluating the suitability of their residential location (Pritchard, 1976; Dennis, 1977; Pooley, 1979).

The residential location decision has been shown to be influenced by a number of factors (Rossi, 1955; Simmons, 1968; Brown and Moore, 1970). In the context of the nineteenth-century city these include constraints of disposable income, the availability of employment in different areas of the city, and access to accommodation in different sectors of the housing market. Social, demographic and cultural influences related to the personal characteristics of the home-seeker must also be considered together with factors such as chance, personal whim and the size of the search area which is itself constrained by an individual's knowledge of the city. Thus at the individual level, the formation of residential areas in Victorian towns was the result of a process of residential decision-making set within a series of choices and con-straints imposed by the characteristics of the city, the society and the individual.

Discussion of residential choice at the

behavioural level poses many problems for the historical geographer, as information on individual decision-making is almost unobtainable for the nineteenth century. The decision-making process of the individual cannot, however, be totally divorced from the aggregate of decisions that have gone before and which will be reflected in existing urban structure. Any decision about residential location is severely constrained by pre-existing conditions, and in this sense it is likely that individual decisions tended to reinforce the *status quo* of existing residential differentiation. When a rare major discontinuity did occur (such as a massive influx of low-status migrants) it temporarily disrupted the residential structure of the nineteenth-century city and caused a major change in the criteria by which residential areas were appraised. Although individual decisions on residential location were thus the basic process causing distinct social areas to develop and retain their character over time, this decision-making process cannot be analysed in isolation. Perhaps of greater importance at any instance in time, and certainly rather easier to analyse, is the pre-existing structure of the city which inevitably influenced an individual's choice, together with the constraints imposed on that individual by the physical structure of the city and the aggregate social, economic and demographic structure of the prevailing society.

A third process to be considered in any appraisal of residential differentiation is the development of commercial and industrial areas within the city which, both directly and indirectly, imposed a number of constraints on the nature of residential development in nineteenth-century towns. These constraints were exerted through pressure on land in certain areas, especially the invasion of inner-city housing by commercial development; through the provision of employment in particular areas with associated working-class housing close by; and, through industrial pollution which caused environmental deterioration in certain areas, thus encouraging those able to move elsewhere to do so. The study of residential areas cannot therefore be divorced from the processes of commercial and industrial development, although precise links are more difficult to formulate. Certainly, the extent to which increased commercial and industrial development actively encouraged residential differentiation is a theme which requires much more detailed investigation.

Fourthly, although land-use planning as such

207

did not exist in the Victorian city and institutional forces were probably less important in the nineteenth century than they are today (Vance, 1976; Harvey, 1974), national and local government together with other institutions could affect residential development in a number of ways. Successive local bye-laws together with national health and housing legislation imposed constraints on new housing development, while town improvement schemes involving, for instance, improved water supply, sewage disposal, closure of cellars and the demolition of slum property affected existing housing (Gauldie, 1974). The rate at which new houses were built, and their tenure and form, was further constrained by factors such as the pattern of landholding, the nature of associated estate development, the prevailing rate of return on investment, and the operation of early building societies (Chapman, 1971; Cannadine, 1977). In the later nineteenth century philanthropic organizations and local authorities also began to intervene in the provision of working-class housing (Tarn, 1973; Taylor, 1974). Although the direct impact of such institutional involvement was in most cases small, the indirect influence of local and national government housing and sanitary reforms on the development of distinctive residential areas should not be underestimated.

Lastly, residential areas in the Victorian city became differentiated through their popular image. Either from personal experience or through semi-official utterances such as reports by Medical Officers of Health, a particular part of the city could gain a reputation as, for example, being predominantly Irish, having a bad record for fever, or suffering from intense industrial pollution. Other areas would become popularly known as high-status areas and once such images of areas became firmly fixed in the minds of the city's inhabitants, they could be further perpetuated through the locational decisions of individuals and thus affect the future development of social areas in the town.

Continuity and change

Because of constraints imposed by the availability of census enumerators' books, detailed studies of nineteenth-century cities (whether cross-sectional or process-oriented) tend to begin in 1851 and end in 1871. All too often the sheer amount of census data to be handled means that only one or two census years are studied in any depth. Inevitably, this had led to many generalizations about the nineteenth-century city being made on the basis of studies of the mid-Victorian period only. The extent to which such studies also represent the early- and late-nineteenth century town has yet to be properly assessed.

If one accepts, however that most change is evolutionary and that the large industrial city did not suddenly alter at some point in the nineteenth century from a state of residential intermixing to one of total residential differentiation, then it is essential to look both forwards and backwards from the mid-Victorian period. The processes which eventually caused distinct residential areas to emerge in large towns in the mid-Victorian period must have been operating over a much longer period prior to their crystallization in distinct social areas and must have continued to influence residential change well into the twentieth century.

With only a few notable exceptions the large industrial centres of the mid-Victorian period were also important in the eighteenth century and earlier. They had substantial foundations on which to grow, and the incipient social areas reflected in occupational and migrant groupings in the eighteenth-century town (Langton, 1977; Langton and Laxton, 1978) may have developed through the early-nineteenth century into a more broadly-based residential differentiation. As the scale of the city increased so the level of differentiation also grew. In the large industrial town the processes outlined above manifested themselves in distinct social areas by the mid-nineteenth century, while in smaller towns residential differentiation remained incipient for a longer period. The way in which these processes collectively produced residential differentiation over a period of time has yet to be fully studied in any town, and the crucial first four decades of the nineteenth century, when many towns were experiencing their most rapid rates of growth, are particularly neglected.

Theories of residential differentiation

While little attention has been paid to the processes causing residential differentiation in Victorian towns, even less has been focused on the development of theories of residential differentiation which are applicable to the nineteenth century. Studies that have couched analysis within a theoretical framework have drawn almost exclusively on the

neo-classical explanations of traditional ecological
theory (Lawton and Pooley, 1975b; Shaw, 1977; Lewis,
1979). The massive bodies of data usually handled
in nineteenth-century urban research have meant that
theory-generation has been neglected in favour of
data description. Whilst empirical analysis will
always be important in historical research, the
interpretation of such data is undoubtedly enhanced
if analysis is undertaken within a carefully-formu-
lated theoretical framework. Furthermore, the
construction of such theory for the study of
residential differentiation in the nineteenth century
may require the integration of material from a number
of other disciplines and time periods. It is
suggested that a useful starting point may be found
in the theories recently generated to investigate
processes causing residential differentiation in
modern urban areas.

R.J. Johnston (1977, 1978) suggests that there
are three distinct approaches to the study of
residential areas in modern urban geography. The
first, and best established, approach he terms 'neo-
classical-functional description' which is typified
by a search for statistical associations between
areas and their attributes. Explanations are
usually related to variations in land values and the
operation of competition and choice in the urban
environment. Such an approach is exemplified by
the classical ecological theories of urban residential development (Burgess, 1924; Theodorson, 1961)
and is the one which has most usually been adopted
by scholars of the Victorian town. A basic premise
of this approach,and of the theories that it
generates,is that different areas have well-defined
attributes and that individuals compete for the most
desirable areas, seeking to maximize the positive
externalities that these areas offer and minimize
negative externalities. At the same time the outcome of such competition may increase the negative
externalities suffered by the residents of other
areas (Johnston and Herbert, 1976; Smith, 1977).

Johnston and Herbert (1976) suggest three main
externality effects which households seek to
maximize by segregating themselves into different
areas of the city. The first concerns the public
behaviour of neighbours, which should coincide with
their own, thus reinforcing the household's social
position. The second relates to the competition
for status and social advancement which can be
gained by locating in an area where there are others
of sufficient status and influence to assist in the
attainment of social aspirations. The third is the

210

competition for property with the desired value. Although traditional ecological theory is now a rather dated and inappropriate concept in the context of modern urban geography, it may still have relevance to the Victorian city. Externality effects would certainly have operated in the nineteenth-century town, and a theory which examines the processes of residential differentiation within a framework of free-market competition by households for the maximization of certain externalities, is one possible extension of the classical ecological approach to the study of residential areas which may be appropriate for studies of the Victorian city.

Secondly, Johnston (1977) discusses the 'behavioural approach' to the study of urban residential differentiation. This approach sees residential location decision-making not in terms of economic theory and competition but instead focuses on the decisions of individuals and the complex range of factors which affect them. This approach to the study of urban residential differentiation developed from those criticisms of classical ecology which emphasized the importance of social values and the perceived image that an area achieves over time (Firey, 1945). More recent refinements have particularly stressed the extent to which decisions about residential location are related to the world as perceived by each individual household, and have noted that few residential moves are made with the benefit of total knowledge of a range of alternative locations (Wolpert, 1965; Buttimer, 1976). Behavioural theories about the formation of residential areas thus also highlight the role of choice in residential relocation, but choice within the environment as perceived by an individual, not necessarily a rational choice made on economic grounds. As suggested above, individual behaviour in the past is difficult to discover, but behavioural theories about residential differentiation, emphasizing choice within the perceived environment, should not be totally neglected by the urban historian.

Third, and most recent, is the 'institutional approach' to the study of residential areas. Whilst previous theories have stressed the choice of individuals and groups (within certain broad constraints) as being the main factor causing residential differentiation, an approach through the study of the role of institutions stresses the constraints imposed on individual households. In general terms these will be economic and social constraints imposed by the prevailing politico-economic system, but in a modern capitalist society, these operate through institutions such as building societies, estate agents, banks, and local authorities. These 'gatekeepers' control access to different sectors of the housing market while the allocation policies followed by each will reflect the social controls operating in society. Most usually these are manifested through discrimination against particular groups in housing allocation (Johnston, 1977, 1978), and not surprisingly the most active protagonists of these theories often (though not necessarily) couch their arguments in Marxist terms (Harvey, 1973, 1974, 1975b; Gray, 1976; Duncan, 1976; Boddy, 1976).

The themes which such studies develop can also be related to the process of residential differentiation in Victorian cities. Harvey (1975a), for instance, attempts to link the historical process of residential differentiation to social theory, arguing that residential differentiation has been inextricably linked to the formation of an accepted series of classes within society. He suggests a process in which, as society became fragmented into classes, so household residential location became severely constrained both by the ideological constraints of society and by those institutions which were in a position to reinforce societal norms by their control over access to different types of housing. Moreover, he asserts that such a process has been operating to cause residential differentiation in British cities since at least the 1850s. This proposition still requires substantive testing through empirical studies of Victorian cities, but an approach which stresses the unequal nature of society, the conflict between groups as one seeks to dominate another, and the ways in which such conflicts impose further constraints on the less fortunate must be an attractive proposition for the study of nineteenth-century society.

Although these may be viewed as three distinct approaches, each generating their own theories, it is perhaps more valuable to see them as complimentary and often overlapping stances (Timms, 1974). Whether we focus our analysis on competition for desirable areas through ability to pay rent, on the status of social areas as perceived by the individual decision-maker, or on the role of society in forming social values and encouraging residential segregation through class conflict, each may be viewed within a framework of choice and constraint. A few sectors of society may be subject to almost

212

total constraint in residential location; others
will have a wider range of choice, although none
will be totally unconstrained.   The precise choices
and constraints operating in particular Victorian
cities will often be different to those in contemporary towns, but the theoretical frameworks
which have been generated by modern urban research
can provide insights into the formation of residential areas in the nineteenth century. This is
particularly so if elements of all three approaches
are synthesized into a theory which emphasizes the
interaction between constraints and choices in
household location.

Choice and constraint in a mid-Victorian town

It has been demonstrated elsewhere that Liverpool
was residentially differentiated by the mid-
Victorian period (Taylor, 1976; Lawton and Pooley,
1976).   This section investigates some of the
processes which caused residential differentiation
in mid-Victorian Liverpool within a conceptual
framework which emphasizes the interaction of those
elements of choice and constraint which affected
various aspects of residential location.

Housing and Environment

Perhaps the major constraint affecting residential
location decisions is the extent to which housing
is available to different sectors of society. A
city in which the housing stock is divided into
distinct sub-markets with clear spatial associations will rapidly develop socially-differentiated
residential areas.   It can be demonstrated that by
the mid-Victorian period different parts of
Liverpool were clearly associated with different
types of housing (Figure 8.1). This pattern of
housing areas had evolved from the later eighteenth
century and during the first half of the nineteenth
century, with most central property consisting
mainly of the oldest and most dilapidated housing,
whilst newer bye-law terrace housing formed a ring
around the urban periphery.   The situation in
Liverpool, however, was somewhat confused by the
persistence of larger terrace housing, mostly dating
from the late-Georgian period, close to the city
centre.   Much of this retained its high status
throughout the nineteenth century (Taylor, 1970;
Treble, 1971; Chalklin, 1974).
    As the population of Liverpool grew rapidly

213

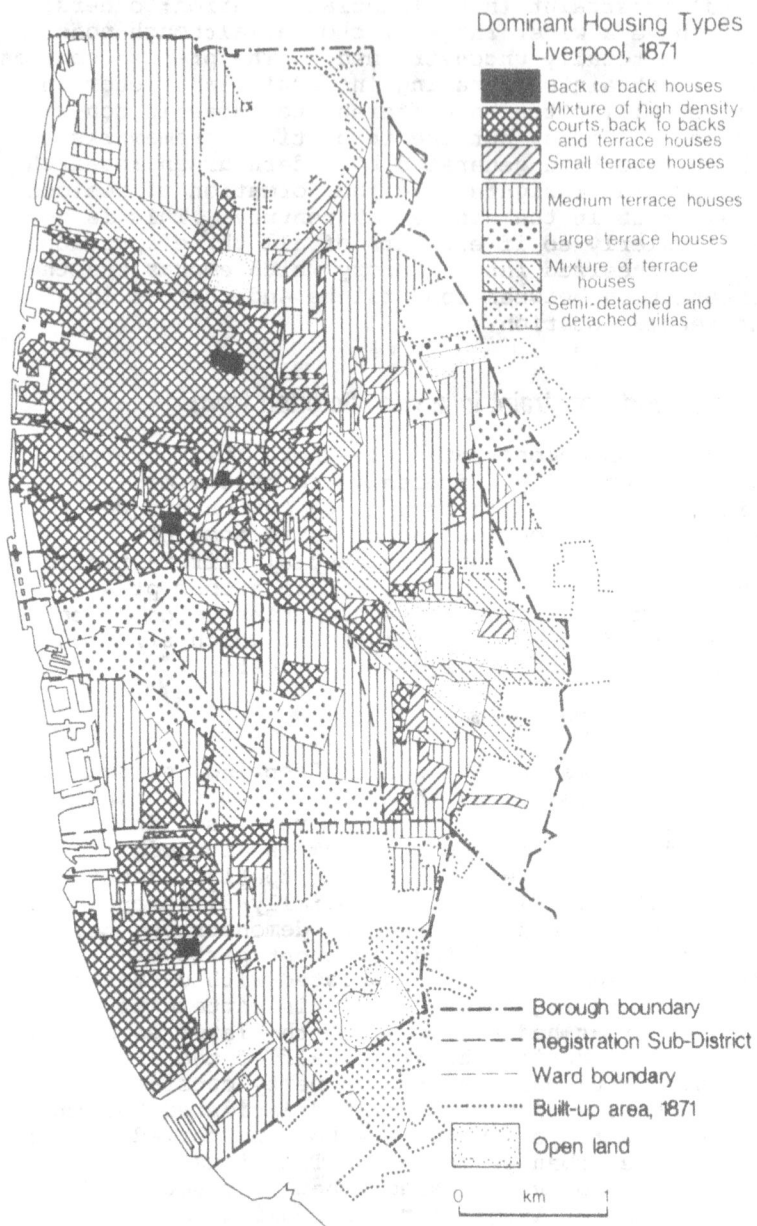

Dominant Housing Types
Liverpool, 1871

- Back to back houses
- Mixture of high density courts, back to backs and terrace houses
- Small terrace houses
- Medium terrace houses
- Large terrace houses
- Mixture of terrace houses
- Semi-detached and detached villas

— · — · — Borough boundary
— — — Registration Sub-District
— — — Ward boundary
·········· Built-up area, 1871
Open land

0   km   1

**Figure 8.1   Distribution of dominant housing types
in Liverpool in 1871**

214

(Table 8.1), housing was made available to the in-
coming population through two related processes
involving the construction of new property and the
subsequent transition of existing property into
lower-status, often multi-occupied, accommodation.
Apart from some small-scale demolition and rebuild-
ing, such as St. Martins Cottages (1869) and the
Nash Grove Scheme in 1885 (Taylor, 1974), there was
little new building in the central area after about
1840 (Table 8.2, Figure 8.2). Not only was new building
spatially concentrated on the urban periphery, but
it also fell predominantly into the middle range of
rentals at £12-£25 per annum (mostly 4s. 6d. - 6s.
per week), thus taking it out of the reach of the
poorest working-class families (Liverpool Council
Proceedings 1866/67; Parry-Lewis, 1965; Laxton,
1970).   Those at the foot of the housing ladder in
Victorian Liverpool thus relied on the effective
operation of the filtering process in order to gain
access to any reasonable accommodation.   As present-
day studies have shown, however, families rarely
filter through housing chains in accordance with
classic ecological theory (Murie, Niner and Watson,
1976; Boddy and Gray, 1979), while those in most
severe need often experience the greatest difficulty
in finding reasonable accommodation.   Although the
topic deserves more detailed research, it can be
shown that over periods of ten or twenty years most
parts of Liverpool exhibited considerable stability
of population characteristics despite very high
rates of residential mobility (Pooley, 1979).   This
suggests that although moving house was easy,
property was not filtering rapidly downwards to
fulfil the demand for low-rent accommodation.   Such
an interpretation is further supported by the
relatively low rate of out-mobility towards the
suburbs that was observed in Liverpool between 1851
and 1881 (Pooley, 1979).   It would thus seem that
much of the new and existing housing stock outside
the older core was being successfully occupied by
in-migrants and newly-formed local households who
could afford rentals of 5s. per week, while increasing demand for the lowest-rent accommodation
had to be met through increased levels of multi-
occupation of the existing housing stock in the
urban core.   The only parts of Liverpool where any
large-scale filtering occurred were in some of the
older Georgian areas, most susceptible to invasion
by lower-status households because of their location (such as Shaw Street on the boundary of
Everton and St. Anne's ward).   These, and much of
the remainder of the existing central housing stock,

215

Table 8.1: Population Change on North Merseyside, 1801-1911

Year	Liverpool Borough	Liverpool Registration District	Toxteth Park[1]	West Derby Registration District[2]
1801	77,653	77,653	2,069	9,925
1811	94,376	94,376	5,864	13,208
1821	118,972	118,972	12,829	19,530
1831	165,175	165,175	24,067	28,991
1841	286,487*	223,003	41,295	47,385
1851	375,955	258,236	61,334	91,945
1861	443,938	269,742	69,284	156,561
1871	493,405	238,411	85,842	257,008
1881	552,508	210,164	117,028	359,275
1891	517,980	156,981	128,387	444,403
1901	684,958*	147,407	136,230	529,684
1911	746,421*	128,673	136,140	613,374

* Indicates boundary changes in preceding decade

[1] Registration sub-district to 1871; registration district from 1881

[2] Excluding Toxteth Park

Source: Census of England and Wales, 1801-1911

were let as multi-occupied accommodation to the mass of the working population, particularly the Irish migrants who entered Liverpool in such large numbers after 1847 (Lawton, 1959; Pooley, 1977). The failure of this supply to match demand is amply demonstrated in successive reports by Medical Officers of Health, who described the grossly-overcrowded conditions and commented frequently on the problems of licensing and controlling multi-occupied lodging houses in the central area and along the dockside.

As local authority housing made little impact in Liverpool (or elsewhere) until after 1918 (Taylor, 1974), and privately-owned accommodation was restricted to a small minority of the population throughout the nineteenth century, it might be suggested that Victorian Liverpool consisted of only one housing market (the rented sector), and thus theories which emphasize differential access to different layers of the housing market would have little relevance for the nineteenth-century city. Access to housing would in these circumstances be governed by ability to pay rent, and those with sufficient income would automatically move to the apparently undifferentiated bye-law terraces whilst the poor would remain in the equally homogeneous inner-city courts. However, modern studies have

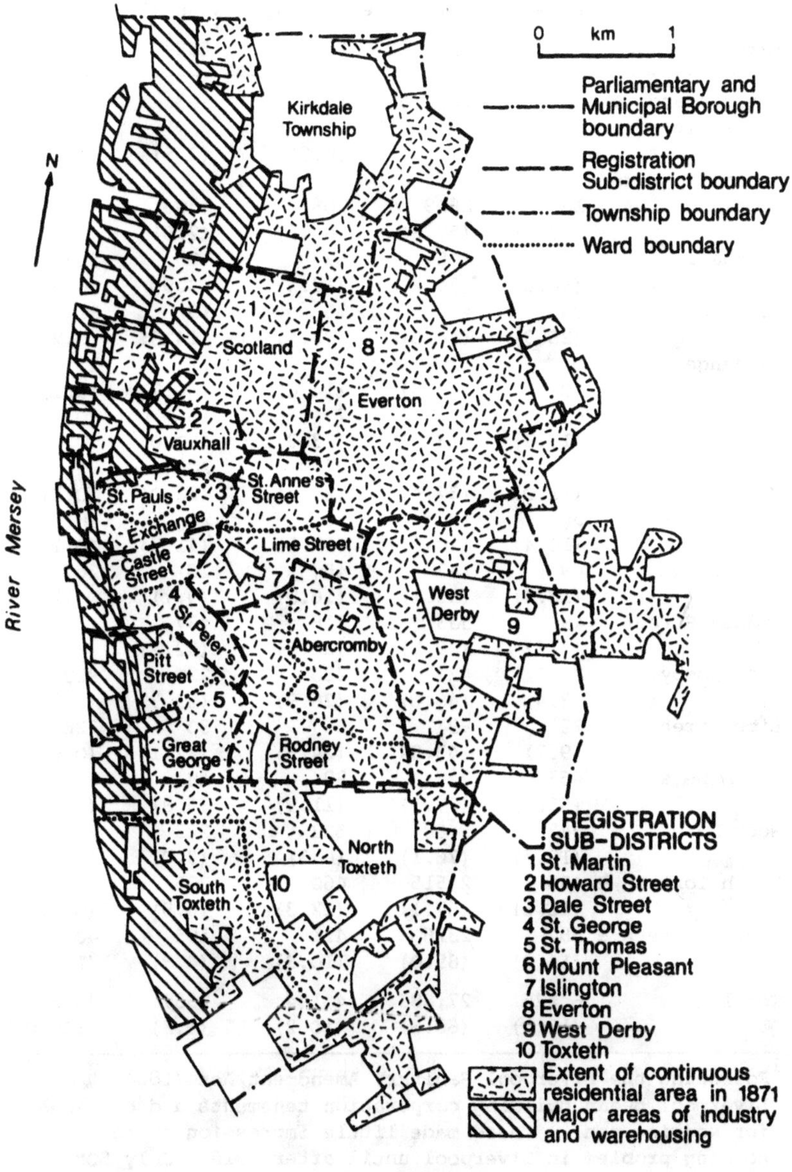
0 km 1
Parliamentary and Municipal Borough boundary
Registration Sub-district boundary
Township boundary
Ward boundary
Kirkdale Township
N
River Mersey
1
Scotland
8
Everton
2
Vauxhall
St. Anne's Street
3
St. Pauls
Exchange
Lime Street
Castle Street
7
4
St. Peter's
West Derby
9
Abercromby
Pitt Street
5
6
Great George
Rodney Street
North Toxteth
South Toxteth
10
REGISTRATION SUB-DISTRICTS
1 St. Martin
2 Howard Street
3 Dale Street
4 St. George
5 St. Thomas
6 Mount Pleasant
7 Islington
8 Everton
9 West Derby
10 Toxteth
Extent of continuous residential area in 1871
Major areas of industry and warehousing

Figure 8.2 Registration sub-districts and wards in Liverpool

Table 8.2:   House building in Liverpool, 1841-1866

Houses built at rentals (per annum)[1]:

Ward	<£12	£12-£25	£25-£35	£35+	Total
Everton and ǂ Kirkdale	1570	12,615	1445	563	16,223
Per cent of total in ward	(9.7)	(77.8)	(8.9)	(3.5)	(39.9)*
Scotland	1885	2609	309	138	4941
	(38.2)	(52.8)	(6.25)	(2.8)	(12.2)
Vauxhall	137	105	53	12	307
	(44.6)	(34.2)	(17.3)	(3.9)	(0.8)
St. Paul	23	40	15	13	91
	(25.3)	(44.0)	(16.5)	(14.3)	(0.2)
Exchange	0	12	4	11	27
	(0)	(44.4)	(14.8)	(40.7)	(0.1)
Castle Street	4	4	7	15	30
	(13.3)	(13.3)	(23.3)	(50.0)	(0.1)
St. Peter's	10	34	27	20	91
	(11.0)	(37.4)	(29.7)	(22.0)	(0.2)
Pitt Street	5	10	13	20	48
	(10.4)	(20.8)	(27.1)	(41.7)	(0.1)
St. George	49	147	45	26	267
	(18.4)	(55.1)	(16.9)	(9.7)	0.7)
Rodney Street	49	533	318	437	1337
	(3.7)	(39.9)	(23.8)	(32.7)	(3.3)
Abercromby	32	427	250	220	929
	(3.4)	(46.0)	(26.9)	(23.7)	(2.3)
Lime Street	22	99	55	50	226
	(9.7)	(43.8)	(24.3)	(22.1)	(0.6)
St. Anne's	457	227	99	48	831
	(55.0)	(27.3)	(11.9)	(5.8)	(2.0)
West Derby ǂ	942	5,683	623	223	7,471
	(12.6)	(76.1)	(8.3)	(3.0)	(18.4)
North Toxteth ǂ	436	2,515	460	335	3746
	(11.6)	(67.1)	(12.3)	(8.9)	(9.2)
South Toxteth ǂ	654	2826	421	195	4096
	(16.0)	(69.0)	(10.3)	(4.8)	(10.1)
Total	6,275	27,885	4,144	2357	40,661
	(15.4)	(68.6)	(10.2)	(5.8)	(100.0)

[1] Following the Liverpool Sanitary Amendment Act (1864)
Liverpool began building corporation tenements and cottages
for working men.   These made little impression on the
housing problem in Liverpool until after 1918. Only 508
dwellings were completed 1869-1891 and 2,895 corporation
dwellings were errected prior to 1916. (Liverpool
Corporation Housing Committee, *Housing progress 1864-1951*,
Liverpool,1951).
See also Parry Lewis (1963) pp.334-339 and Treble (1971) for

Table 8.2 (cont'd)
    further details of housebuilding in Liverpool. On the contri-
    bution of corporation housing to the working-class housing
    problem in Liverpool see Taylor (1974).
‡Wards outside Liverpool Parish experiencing most rapid
    suburban development.
*Percent of total houses built in Liverpool 1841-1866.
    Source: Liverpool Council Proceedings, 1866/67.

shown that many factors can affect the desirability
of a particular house or area (Bourne and Hitchcock,
1979), and that an apparently homogeneous housing
market may in fact be divided into a number of sub-
markets each with distinct spatial associations
(Harvey, 1974).   As well as the crude distinctions
that can be made between back-to-back houses, high-
density courts and terraces, larger terrace houses
and semi-detached villas (Figure 8.1),there was a much
more subtle differentiation within the large
privately-rented sector of mid-Victorian Liverpool.
      For instance, even within the central zone of
high-density court and terrace accommodation which
provided for those at the lowest end of the housing
ladder, a number of important 'sub-markets' can be
discovered.   This differentiation is best illustrated through the comments of contemporaries who
observed these subtle distinctions.   One of the most
constant hazards of Victorian life was exposure to
infectious disease.   Whilst mortality statistics at
ward level demonstrated a simple relationship between
high-density living in the central core and deaths
due to infectious diseases (Duncan, 1843), health
reports also emphasized that fever and cholera struck
with most devastating effect in only a few restricted
areas: 'Cholera in its epidemic virulence was restricted to the lowest, dirtiest, most crowded and
most squalid streets of the Borough' (Trench, 1866,
p.29).   Other reports isolated a few small areas
where fever was endemic, such as the block of
streets in Exchange Ward described as follows in
1843: ⟋It has⟋'...the largest amount of cellar
population, the greatest number of damp cellars, the
courts are of the worst construction, the sewerage
is more defective than in any part of the town...few
of the front houses have privies or ashpits, most of
the streets and courts are the filthiest in the town,
and density of population here reaches its highest
point' (Duncan, 1843, p.54).   Similarly, some parts
of the large area of high-density court and terrace
housing suffered more severely from industrial
pollution than others, so that even within this zone
of worst housing some small areas took on a better

219

reputation than others. Census evidence confirms
that there were small but none the less significant
differences in the characteristics of the population
of almost adjacent courts (Pooley, 1978), reflecting
the externality effects which operated between areas
within one broad housing class.

It would appear, however, that rents were
mainly dependent on the number of rooms let, with a
three-roomed front house costing 4s. 6d. - 5s. per
week in the 1870s, a three-roomed court house
2s. 6d. - 3s., and a single room 1s. 3d. - 2s. per
week (Trench, 1875, pp.43-7). It should also be
noted that rents in the central area were often not
significantly less than for a new terrace house in
Everton or Kirkdale, which could be rented for as
little as 4s. 6d. - 6s. per week (Liverpool Council
Proceedings 1866/7; Lawton and Pooley, 1975a). It
thus seems that cost was not the only factor controlling access to housing and it is pertinent to
enquire more deeply into the constraints causing
society to divide itself between these different
housing classes.

In many cases economic factors clearly played
an important part. For the lowest paid, the unemployed, widows left to cope with a large family
and other equally disadvantaged classes, 1s. 6d. per
week for a single room in an inner-city court was
genuinely all that could be afforded. Contemporaries, too, emphasized economic constraints, but
often with certain qualifications. Examine, for
example, this comment on the relatively high rents
being charged for rooms in a new block of labourers'
cottages in the 1860s: 'Any man, more particularly
an English man, who can pay 6s. per week, will
locate himself in a neat cottage house, where he
can have that independent action and domestic
privacy which, to some extent, cannot be obtained in
large blocks of dwellings with a common entrance'
(Shimmin, 1862/3, pp.388-89). Differentiation was
thus clearly seen to be the result of variable
ability to pay rent, but this could be qualified
both by migrant origin and, as will be demonstrated
below, by the nature of a person's employment.

Furthermore, although institutional control of
the housing market was much less in the nineteenth
century than it is today, there were other ways in
which constraints could be imposed and access denied
to better-quality terrace housing or more respectable courts. In the newer terraced suburbs landlords tended to live in the area in which they owned
other houses. They thus had a vested interest in
maintaining the character of the suburb and were in

220

a position to impose their views regarding the suitability of a tenant for the neighbourhood. Even in the poorer areas other tenants and property owners could exercise control over the acceptability of particular tenants, as evidenced by the comment of the 'Land and House Owners Association' that '...the cleanly poor would insist on a dirty person removing from a court, and the owners of property were obliged to compel them to leave for fear of losing their other tenants' (Parkes and Sanderson, 1871). Few landlords actually lived in the high-density courts of central Liverpool, and many were complete absentees who simply sent a weekly rent collector and scarcely fulfilled their legal obligations on maintenance and the provision of basic sanitary facilities (Shimmin, 1862/3). Subtle differences between areas and courts were thus most often the result of social control operating through a consensus of local opinion.

Prejudice against certain groups in society, particularly the Irish, was common in the nineteenth century and was often most vehemently articulated by influential men in public positions (Duncan, 1859; Trench, 1866). These prejudices could easily be translated into discrimination, either through non-Irish residents effectively excluding Irish migrants, or through landlords steering people into the accommodation they considered most appropriate for their needs, in much the same way as a building society, estate agency or local authority may do today (Duncan, 1976; Gray, 1976). It would have been easy to restrict access to better-class terrace housing or certain courts by charging differential rents, asking for references, demanding a substantial deposit or by seeking payment of rent in advance. Although such practices are difficult to prove in an historical context there is ample contemporary comment to suggest that they did occur, and there is no reason to assume that, because formal institutions were few, there were no 'gatekeepers' restricting access to sub-classes within the housing market. As Harvey (1974) suggests, in a situation of rapidly increasing population (such as the nineteenth-century city) a form of class-monopoly rent can easily develop and the hold of the landlord over tenant can be increased.

Access to the owner-occupied sector of the housing market was very restricted in the nineteenth century. The extent to which middle-income and working-class families had the opportunity for owner-occupancy depended mainly on the activities

of early building societies, but could also relate
to the patterns of land-ownership and building
development. Most early building societies were
of the terminating variety which were wound up when
all subscribers had achieved home ownership, but the
impact that they had on housing provision varied
greatly from town to town (Cleary, 1965; Chapman,
1971; Robson, 1973; Boddy, 1980). For industrial
towns a figure of ten per cent owner-occupancy would
have been high (Daunton, 1977), and although the
absence of rate books means that a simple calculation of owner-occupancy rates cannot be made for
mid-Victorian Liverpool, it is likely that owner-
occupiers formed rather less than ten per cent of
the Merseyside population. Liverpool certainly had
building societies during the nineteenth century,
but the financial demands of home-ownership usually
restricted membership to households with a substantial regular income. Terminating societies in the
1840s and 50s could demand subscriptions of as much
as 10 shillings per month (Robson, 1973), effectively excluding most working-class Liverpudlians.
Victorian Liverpool thus developed a housing
stock which was spatially differentiated in terms
of its physical characteristics, and was further
differentiated by the varying quality of the
environment in different areas of the city. Access
to different types of housing was controlled not
only by economic factors but also by the influence
of landlords, building societies and other tenants.

Work

Not only was the newcomer to the Victorian city involved in the search for accommodation, but he also
needed to search for employment. The type of work
in which the principal wage-earner was engaged
restricted a household's residential location in
two main ways. First, the level and the
regularity of income affected the extent to which a
household could commit itself to a substantial
weekly outlay on rent or to a building society.
Second, the location of the workplace and the
importance of a short journey to work placed further
constraints on residential location. In Liverpool,
mass transport for the working man did not develop
until the late-nineteenth century (Monroe, 1967;
Dyos and Aldcroft, 1969); throughout the period
when residential differentiation was developing
Liverpool was a walking city for all but the very
rich.
For some occupations a considerable journey to

222

work did not matter. Clerks, for instance, lived predominantly in the newer suburbs of Everton and West Derby, thus entailing a journey to work of 1½ miles or more. On the other hand, those employed in dock work clustered around the dockside and rarely lived more than a mile from a potential work-place (Orchard, 1871; Lawton and Pooley, 1976; Anderson, 1976). The contrasting residential dis-tributions of these two occupational groups reflec-ted the difference between a job which had regular and certain employment, and work on the docks which was casual in nature and where proximity to a hire stand was essential in order to secure work when a ship came in (Rathbone, 1904; Williams, 1912).

Although there were undoubtedly important differences in status and aspiration between dock workers and clerks which also affected their residential choices, it remains true that many of the more skilled dock workers in regular employment could easily have afforded the rent of a decent terraced cottage in the suburbs. Their dilemma was that if they moved from overcrowded, sub-standard, dockside accommodation the chances of gaining regular employment would recede and their income would fall. The problem was acknowledged by con-temporary commentators who noted the way in which families with reasonable incomes were forced to live in filth and squalor in dockside courts and fre-quently 'the convenience of his home to his labour was given as the reason for remaining in such a doghole' (Shimmin, 1862/3 p.268).

Thus men in different occupations, even if they had a similar income and social standing, did not necessarily have equal access to a range of housing in the city. The casual dock worker in particular was severely constrained by the need to live close to his place of work, and this factor could completely override any freedom of choice economic circumstances might otherwise have permitted. Not surprisingly, the better-off dockworker who was seeking a reasonable home life tried to live in those areas of inner-city housing least affected by the negative externalities discussed above. Such clustering further accentuated small-scale differen-tiation within this housing class.

Migrant origin

It can also be demonstrated that residential clustering in Victorian Liverpool developed because of the origin of a migrant household. This is particularly true of those migrants, such as the

Irish, Welsh, and to a lesser extent Scots and over-seas-born, who had particularly strong cultural links with their homeland or who came from a mainly rural background, and thus were not easily assimilated to urban life. Where a group was also at an economic disadvantage because of the circumstances of its movement (such as Irish famine migrants), the tendency towards residential clustering was even greater (Pooley, 1977).

The Irish had settled in north Liverpool in the early-nineteenth century, forming a migrant colony which gave them security in a strange environment and sheltered them from the prejudice and discrimination they faced (P.P.XXXIV, 1836,pp.8-41). The area in which they settled rapidly came under severe pressure from industrial and commercial development as the city expanded. Despite a deteriorating environment and housing stock, migrants from Ireland continued to be constrained by the cultural barriers and discrimination they faced in Liverpool which, together with economic circumstances, forced them to remain clustered in the same area of the city. This placed increasing pressure on an almost static housing stock, forced out those more res-pectable English residents who were able to move, and caused further deterioration of housing conditions in the inner city.

Other migrants, especially the Welsh, were also attracted to particular migrant enclaves which developed in Liverpool, thus adding a further dimension to the developing residential differentiation (Pooley, 1977). The decision to locate in a migrant area, especially in the case of Welsh or Scots migrants, was partly one of choice, but it was also imposed by the cultural dissonance that many newly-arrived migrants faced. It is likely that many migrants, feeling constrained to go straight to a migrant area, found themselves in accommodation which on other counts they might have deemed un-suitable.

Individual characteristics and aspirations

Lastly, residential differentiation developed as a result of the varied household characteristics and individual aspirations of Liverpool's population. Life-cycle stage is an important factor which stimulates mobility (Rossi, 1955; Pooley, 1979) and households with different types of housing demand will locate in different areas of a city. In part, this reflects a degree of choice in residential location, but such households are also constrained

224

by the availability of different types of housing in different areas: the single male migrant seeking lodgings would almost inevitably have found himself in the area of dockside lodging houses in Liverpool.

The effects of individual aspirations and perceptions of status on residential differentiation are difficult to measure in the past, but contemporaries were well aware of these differences, especially as they pertained to the broad range of skilled working men with reasonable incomes. Commentators frequently blamed bad living conditions on the uncleanly habits and intemperance of court inhabitants (Shimmin, 1862/3; Parkes and Sanderson, 1871), but were quick to distinguish between the intemperant and the provident working man who saved regularly, drank little and kept a respectable home (Shimmin, 1860/61). Although constraints of poverty and the other controls discussed above were probably more important than individual habits in condemning a family to a life-time in the courts, it is undoubtedly true that a working-class family with sufficient will and aspiration could extricate itself from the worst housing and maintain a decent home in the terraced suburbs (Lawton and Pooley, 1975a).

Conclusions

This paper has only selectively illustrated some of the processes through which residentially-differentiated social areas emerged in the mid-Victorian city. Interaction of the different forces was complex, but collectively they provided a sorting mechanism which, though operating on different spatial and temporal scales, affected all towns during the nineteenth century.

A conceptual framework which views urban residential location as the outcome of conflicting choices and constraints illustrates the way in which this sorting mechanism operated in different strata of Victorian society. Victorian Liverpool was an unequal city in which the poorest were clearly the most restricted in their choice of housing, not just by their low income but often by the nature of their work, their migrant origins and by outright discrimination. Even among the better-off, few residential location decisions were completely unconstrained. As economic constraints lessened other, more subtle, forces came into play. For the rich, constraints imposed by aspirations for social advancement and the resulting need to locate in a socially-respectable area may have effectively reduced choice as much as poverty did for other families.

The resultant residential differentiation produced by these forces manifested itself most clearly at the extremes of society, but it was not only the very rich and the very poor who lived apart. Different working-class areas of Victorian Liverpool were viewed by contemporaries as reflecting varying levels of respectability and, at the micro-scale, whilst one inner-city court may have consisted predominantly of English dockers' families in decent homes another may have been wholly Irish with a quite different reputation. Certainly, there were opportunities for social as well as residential mobility in the Victorian city and we must beware of static stereotypes of the different segments of Victorian society. Many higher-status Irish were able to move into predominantly non-Irish areas; some unskilled or semi-skilled dock workers did live in zones of predominantly skilled working-class housing. These, however, were probably families that conformed to the social norms of the areas to which they were moving, and as such reinforced rather than reduced the distinctive social differences between areas of the town.

Although this essay has merely summarized part of a study of Liverpool and suggested pointers for future research on nineteenth-century urban residential differentiation, there are some more general conclusions that may be drawn. It is no longer sufficient to describe the main dimensions of Victorian urban structure or simply to map the residential distributions of social classes and migrant groups. Although these are important preliminary exercises, it is necessary to expand such work into an investigation of the processes which caused residential segregation to develop in Victorian towns. Such investigations should go beyond the bald statement of social and economic differences contained in census occupational statistics by assessing the complex economic and social forces which affected individual residential location. Furthermore, explanations should be set within a clearly-stated theoretical framework. One approach, drawing on a synthesis of current urban geographical theory, is to view the residential location decision as the outcome of the interaction of specific choices and constraints within Victorian society. Research which focuses on the processes of residential change, rather than the resulting patterns of residential distribution, will

eventually allow much fuller explanations of the
nature of residential differentiation and the implications which this had for life in an essentially
inegalitarian urban environment.

References

Anderson, G.L.P. (1976) *Victorian clerks,* Manchester
University Press, Manchester

Armstrong, W.A. (1974) *Stability and change in an
English county town. A social study of York 1801-51,*
Cambridge University Press, Cambridge

Berry, B.J.L. (ed.)(1971) 'Comparative factorial
ecology', *Economic Geography,* 47, (Supplement)

Berry, B.J.L. and Rees, P.H. (1969) 'The factorial
ecology of Calcutta', *American Journal of Sociology,* 74, 445-
91

Berry, B.J.L. and Smith, K.B. (eds.) (1972) *City
classification handbook: methods and applications,* Wiley,
New York

Boddy, M. (1980) *The Building Societies,* Macmillan,
London

Boddy, M. and Gray, F. (1979) 'Filtering theory, housing
policy and the legitimation of inequality', *Policy and
Politics,* 7, 39-54

Bourne, L.S. and Hitchcock, J.R.(eds.) (1979)*Urban housing
markets: recent directions in research and policy,* University
of Toronto Press, Toronto

Bowley, A.L. and Burnett-Hurst, A.R. (1915) *Livelihood
and poverty,* Bell, London

Brown, L.A. and Moore, E.G. (1970) 'The intra-urban
migration process: a perspective', *Geografiska Annaler,* 52B,
1-13

Burgess, E.W. (1924) 'The growth of the city: an introduction to a research project', *Proceedings and Papers of the
American Sociology Society,* 18, 85-97

Burnett, J. (1978) *A social history of housing,* David
and Charles, Newton Abbot

Buttimer, A. (1976) 'Grasping the dynamism of life
world', *Annals of the Association of American Geographers,*
66, 277-92

Cannadine, D. (1977) 'Victorian cities. How different?',
Social History, 4, 457-82

Chalklin, C. (1974) *The provincial towns of Georgian
England,* Arnold, London

Chapman, S.D. (ed.) (1971) *The history of working-class
housing,* David and Charles, Newton Abbot .

Clark, D., Davies, W.K.D. and Johnston, R.J. (1974)
'The application of factor analysis in human geography',
The Statistician, 23, 259-81

Clark, P. (1972) 'The migrant in Kentish towns, 1580-

1640', in P. Clark and P. Slack (eds.) *Crisis and order in English towns 1500-1700*, Routledge, London, pp.117-63

Cleary, E.J. (1965) *The building society movement*, Elek Books, London

Daunton, M.J. (1977) *Coal metropolis: Cardiff, 1870-1914*, Leicester University Press, Leicester

Davies, W.K.D. and Lewis, G.J. (1973) 'The urban dimensions of Leicester, England', in B.D. Clark and M.B. Gleave (eds.), 'Social patterns in cities', *Institute of British Geographers. Special publication*,5, pp.71-86

Dennis, R.J. (1977) 'Intercensal mobility in a Victorian city', *Transactions of the Institute of British Geographers*, N S 2, 349-63

Dogan, M. and Rokkan, S. (eds.) (1969) *Quantitative ecological analysis in the social sciences*, M.I.T. Press, Cambridge, Mass.

Duncan, S.S. (1976) 'Research directions in social geography: housing opportunities and constraints', *Transactions of the Institute of British Geographers*, N S 1, 10-19

Duncan, W.H. (1843) *On the physical causes of the high rate of mortality in Liverpool*, Liverpool

Duncan, W.H. (1859) *Report on the health of Liverpool during the year 1859*, Liverpool

Dyos, H.J. (1967) 'The slums of Victorian London', *Victorian Studies*, 11, 5-40

Dyos, H.J. and Aldcroft, D.H. (1969) *British transport. An economic survey from the seventeenth century to the twentieth*, Leicester University Press, Leicester

Engels, F. (1845, 1969 edition) *The condition of the working class in England*, Panther, London

Firey, W. (1945) 'Sentiment and symbolism as ecological variables', *American Sociological Review*, 10, 140-8

Flinn, M.W. (ed.) (1965) *Edwin Chadwick's report on the sanitary conditions of the labouring population of Great Britain 1842*, Edinburgh University Press, Edinburgh

Friedlander, D. (1971) 'The spread of urbanization in England and Wales 1851-1951', *Population Studies,* 24, 423-43

Gauldie, E. (1974) *Cruel habitations. A history of working-class housing 1780-1918*, Allen and Unwin, London

Goheen, P.G. (1970) *Victorian Toronto 1850-1900: pattern and process of growth,* University of Chicago, Chicago

Gray, F. (1976) 'Selection and allocation in council housing', *Transactions of the Institute of British Geographers*, N S 1, 34-46

Harris, C.D. and Ullman, E.L. (1945) 'The nature of cities', *Annals of the American Academy of Political and Social Science,* 242, 7-17

Harvey, D. (1973) *Social justice and the city*, Arnold, London

Harvey, D. (1974) 'Class-monopoly rent, finance capital, and the urban revolution', *Regional Studies*, 3, 239-55

Harvey, D. (1975a) 'Class structure in a capitalist

228

society and the theory of residential differentiation', in
R. Peel, M. Chisholm and P. Haggett (eds.), *Processes in
physical and human geography: Bristol essays*, Heinemann,
London, pp.354-72

Harvey, D. (1975b) 'The political economy of urbanization in advanced capitalist societies: the case of the United
States', in G. Gappert and H.M. Rose (eds.), *The social
economy of cities*, Sage, Beverley Hills, pp.119-64

Hill, F. (1975) *Victorian Lincoln*, Cambridge University
Press, Cambridge

Hoyt, H. (1939) *The structure and growth of residential
neighbourhoods in American cities*, Federal Housing Administration, Washington, D.C.

Hunter, A.A. (1972) 'Factorial ecology: a critique and
some suggestions', *Demography*, 9, 107-17

Johnston, R.J. (1971) *Urban residential patterns*, Bell,
London

Johnston, R.J. (1972) 'Towards a general model of intra-
urban residential patterns: some cross-cultural descriptions',
Progress in Geography, 4, 83-124

Johnston, R.J. (1976) 'Residential area characteristics:
research methods for identifying urban sub-areas - social area
analysis and factorial ecology', in D.T. Herbert and
R.J. Johnston (eds.) *Social areas in cities Volume 1*, Wiley,
London, pp.193-235

Johnston, R.J. (1977) 'Urban geography: city structures',
Progress in Human Geography, 1, 118-129

Johnston, R.J. (1978) 'Urban geography: city structures',
Progress in Human Geography, 2, 148-152

Johnston, R.J. and Herbert, D.T. (1976) 'An introduc-
tion: spatial processes and form', in D.T. Herbert and
R.J. Johnston (eds.), *Social areas in cities, volume 1.
Spatial processes and form*, Wiley, London, pp.5-18

Keene, D.J. (1979) 'Medieval Winchester: its spatial
organization' in B.C. Burnham and J. Kingsbury (eds.) *Space,
Hierarchy and Society. Interdisciplinary studies in social
area analysis*, B.A.R. International series 59, Oxford,
pp.149-160

Kellett, J. (1969) *The impact of railways on Victorian
cities*, Routledge, London

Langton, J. (1975) 'Residential patterns in pre-
industrial cities: some case studies from seventeenth-century
Britain', *Transactions of the Institute of British
Geographers*, 65, 1-28

Langton, J. (1977) 'Late medieval Gloucester: some data
from a rental of 1455' *Transactions of the Institute of
British Geographers*, N S 2, 259-77

Langton, J. and Laxton, P. (1978) 'Parish registers and
urban structure. The example of late-eighteenth century
Liverpool', *Urban History Yearbook*, pp.74-84

Law, C.M. (1967) 'The growth of urban population in
England and Wales 1801-1911' *Transactions of the Institute*

of British Geographers, 41, 125-43

Lawton, R. (1959) 'Irish migration to England and Wales in the mid-nineteenth century', *Irish Geography,* 4, 35-54

Lawton, R. (1967) 'Rural depopulation in nineteenth-century England', in R.W. Steel and R. Lawton (eds.) *Liverpool Essays in Geography. A jubilee collection,* Longmans, London, pp.227-256

Lawton, R. (ed.) (1978) *The census and social structure. An interpretative guide to nineteenth-century censuses for England and Wales,* Cass, London

Lawton, R. and Pooley, C.G. (1975a) 'Individual appraisals of nineteenth-century Liverpool', *Social geography of nineteenth-century Merseyside project. Working paper,* 3

Lawton, R. and Pooley, C.G. (1975b) 'The urban dimensions of nineteenth-century Liverpool', *Social geography of nineteenth-century Merseyside project. Working paper,* 4

Lawton, R. and Pooley, C.G. (1976) *The social geography of Merseyside in the nineteenth century,* Final Report to the S.S.R.C.

Laxton, P. (1970) 'The built-up area, 1800-1913', in J.A. Patmore and A.G. Hodgkiss (eds.) *Merseyside in maps,* Longmans, London, p.17

Lewis, C.R. (1979) 'A stage in the development of the industrial town: a case study of Cardiff, 1845-75', *Transactions of the Institute of British Geographers,* N S 4, 129-152

Liverpool Borough Council Proceedings, (1866-67)

Mather, P.M. and Openshaw, S. (1974) 'Multivariate methods and geographical data', *The Statistician,* 23, 283-308

Monroe, S.A. (1967) 'Tramway companies in Liverpool 1859-1897', *Transactions of the Historic Society of Lancashire and Cheshire,* 119, 181-212

Murdie, R.A. (1969) *Factorial ecology of metropolitan Toronto 1951-61. An essay on the social geography of the city,* University of Chicago, Chicago

Murdie, R.A. (1976) 'Spatial form in the residential mosaic', in D.T. Herbert and R.J. Johnston (eds.) *Social Areas in Cities. Volume 1. Spatial processes and form,* Wiley, London, pp.237-72

Murie, A., Niner, P. and Watson, C. (1976) *Housing policy and the housing system,* Allan and Unwin, London

Orchard, B.G. (1871) *The clerks of Liverpool,* Liverpool

Parkes, E.A. and Sanderson, J.B. (1871) *Report on the sanitary condition of Liverpool,* Liverpool

Parliamentary Paper (1836) XXXIV, *Report on the state of the Irish poor in Great Britain*

Parliamentary Paper (1845) XVIII, *Second report of the commissioners for inquiring into the state of large towns and populous districts*

Parry-Lewis, J. (1965) *Building cycles and Britains growth,* Manchester University Press, Manchester

Perkin, H. (1970) *The age of the railway,* Panther, London

Pooley, C.G. (1973) 'Problems of topographical indexing and spatial analysis of small area data for nineteenth-century cities', in R. Lawton and C.G. Pooley (eds.), 'Methodological problems in the statistical analysis of small area data', *Social geography of nineteenth-century Merseyside project. Working paper* 2, pp.43-54

Pooley, C.G. (1977) 'The residential segregation of migrant communities in mid-Victorian Liverpool', *Transactions of the Institute of British Geographers*, N S 2, 364-82

Pooley, C.G. (1978) 'Migration, mobility and residential areas in nineteenth-century Liverpool', unpublished Ph.D. thesis, University of Liverpool

Pooley, C.G. (1979) 'Residential mobility in the Victorian city', *Transactions of the Institute of British Geographers*, N S 4, 258-77

Pritchard, R.M. (1976) *Housing and the spatial structure of the city*, Cambridge University Press, Cambridge

Rathbone, E.F. (1904) *Report of an enquiry into the conditions of dock labour at Liverpool docks*, Liverpool

Rees, P.H. (1970) 'Concepts of social space: towards an urban social geography', in B.J. Berry and F.E. Horton (eds.), *Geographic perspectives on urban systems*, Prentice Hall, Englewood Cliffs, N.J. pp.306-94

Rex, J. and Moore, R. (1967) *Race, community and conflict. A study of Sparkbrook*, Oxford University Press, London

Robson, B.T. (1969) *Urban analysis. A study of city structure*, Cambridge University Press, Cambridge

Robson, B.T. (1973a) 'A view on the urban scene', in M. Chisholm and B. Rodgers (eds.), *Studies in Human Geography*, Heinemann, London, pp.203-41

Robson, B.T. (1973b) *Urban growth. An approach*, Methuen, London

Rossi, P.H. (1955) *Why families move*, Free Press, Glencoe, Ill.

Savanger, J. (1978) 'Pittsburgh's residential pattern in 1815', *Annals of the Association of American Geographers*, 68, 265-77

Shaw, M. (1977) 'The ecology of social change: Wolverhampton 1851-71', *Transactions of the Institute of British Geographers*, N S 2, 332-48

Shevky, E. and Bell, W. (1955) *Social area analysis: theory, illustrative application and computational procedure*, Greenwood Press, Stanford

Shevky, E. and Williams, M. (1949) *The social areas of Los Angeles: analysis and typology*, University of California Press, Los Angeles

Shimmin, H. (1860/61) 'Amusements of the people and conditions of working men', *Porcupine*, 1, 32 (continuing)

Shimmin, H. (1862/63) 'The mysteries of the courts', *Porcupine*, 4, 260 (continuing)

Simmons, J.W. (1968) 'Changing residence in the city',

Geographical Review, 58, 622-51

Sjoberg, G. (1960) *The pre-industrial city: past and present*, Free Press, Glencoe, Ill.

Smith, D.M. (1977) *Human geography: a welfare approach*, Arnold, London

Sweetser, F.L. (1962) *Patterns of change in the social ecology of metropolitan Boston 1950-1960*, Massachusetts Department of Mental Health, Boston, Mass.

Tansey, P.A. (1973) 'Residential patterns in the nineteenth-century city: Kingston-upon-Hull, 1851', unpublished Ph.D. thesis, University of Hull

Taeuber, K.E. and Taeuber, A.F. (1965) *Negroes in cities: residential segregation and neighbourhood change*, Aldine, Chicago

Tarn, J.N. (1973) *Five-per-cent philanthropy: an account of housing in urban areas between 1840 and 1914*, Cambridge University Press, Cambridge

Taylor, I.C. (1970) 'The court and cellar dwelling: the eighteenth-century origin of the Liverpool slum', *Transactions of the Historic Society of Lancashire and Cheshire*, 122, 67-90

Taylor, I.C. (1974) 'The insanitary housing question and tenement dwellings in nineteenth-century Liverpool' in A. Sutcliffe (ed.), *Multi-storey living. The British working-class experience*, Croom Helm, London, pp.41-87

Taylor, I.C. (1976) 'Black Spot on the Mersey', unpublished Ph.D. thesis, University of Liverpool

Theodorson, G.A. (ed.) (1961) *Studies in Human Ecology*, Harper and Row, New York

Timms, D.W.G. (1971) *The urban mosaic: towards a theory of residential differentiation*, Cambridge University Press, Cambridge

Timms, D.W.G. (1976) 'Social bases to social areas', in D.T. Herbert and R.J. Johnston (eds.), *Social areas in cities. Volume 1. Spatial processes and form*, Wiley, London, pp.19-40

Treble, J.H. (1971) 'Liverpool working-class housing 1801-1851', in S.D. Chapman (ed.), *The history of working-class housing*, David and Charles, Newton Abbot, pp.167-220

Trench, W.S. (1866) *Report on the health of Liverpool during the year 1866*, Liverpool

Trench, W.S. (1875) *Report of the medical officer of health as to the area, population and death rate of two districts of the Borough*, Liverpool

Vance, J.E. (1971) 'Land assignment in pre-capitalist, capitalist and post-capitalist cities', *Economic Geography*, 47, 101-20

Vance, J.E. (1976) 'Institutional forces that shape the city', in D.T. Herbert and R.J. Johnston (eds.), *Social areas in cities: Volume 1. Spatial processes and form*, Wiley, London, pp.81-109

Vaughan, R. (1843) *The age of great cities, or modern society viewed in relation to Intelligence, Morals and Religion*, London

232

Ward, D. (1964) 'A comparative historical geography of street-car suburbs in Boston, Mass. and Leeds, England: 1850-1920', *Annals of the Association of American Geographers*, 54, 447-89

Ward, D. (1975) 'Victorian cities: how modern?', *Journal of Historical Geography*, 1, 135-151

Ward, D. (1975) 'The Victorian slum: an enduring myth?' *Annals of the Association of American Geographers*, 66, 323-36

Ward, D. (1980) "Environs and neighbours in the 'Two Nations', residential differentiation in mid-nineteenth century Leeds", *Journal of Historical Geography*, 6, 133-62

Warner, S.B. (1962) *Streetcar suburbs: the process of growth in Boston,* Harvard University Press, Cambridge, Mass.

Warnes, A.M. (1970) 'Early separation of homes from work-places', *Transactions of the Historic Society of Lancashire and Cheshire,* 122, 105-36

Warnes, A.M. (1972) 'Residential patterns in an emerging industrial town', in B.D. Clark and M.B. Gleave (eds.), 'Social patterns in cities', *Institute of British Geographers Special Publication,* 5, 169-89

Weber, A.F. (1899, 1963 edition) *The growth of cities in the nineteenth century,* Cornell University Press, New York

Williams, R. (1912) *The Liverpool Docks Problem,* Liverpool

Wolpert, J. (1965) 'Behavioural aspects of the decision to migrate', *Papers and Proceedings of the Regional Science Association,* 15, 159-69

Woods, R.I. (1976) 'Aspects of the scale problem in the calculation of segregation indices: London and Birmingham 1961-1971', *Tijdschrift voor Economische en Sociale Geographie,* 67, 169-74

Wrigley, E.A. (ed.) (1972) *Nineteenth-century society. Essays in the use of quantitative methods for the study of social data,* Cambridge University Press, Cambridge

Chapter 9

RESIDENTIAL DIFFERENTIATION IN NINETEENTH-CENTURY TOWNS: FROM SHAPES ON THE GROUND TO SHAPES IN SOCIETY

David Cannadine

Introduction

According to a recent editorial of the *Urban History Yearbook* (Dyos, 1978, p.3), 'the basic commitment of the urban historian is to a deliberate and ex-
plicit emphasis on the urban process and the urban presence in the broader history of society', by which is meant both the place itself - the town, the city - and the human constituents - the urban popu-
lation. And, it goes on to argue, since 'the place alone cannot constitute itself' and since 'its human constituents cannot hive themselves off from it', the fundamental question is, 'how...do they connect?' Nor is this a question to which urban historians are alone in addressing themselves. For the *Yearbook's* question - consciously or un-
consciously - echoes the argument made by David Harvey (1975, p.23) who suggested that geographers, in evolving general theories of the city, 'must somehow relate the social processes of the city to the spatial form which the city assumes'. But still the question remains: what, if anything, theoretically or empirically, are the links between the shapes on the ground - the physical form which the evolving city took - and the shapes in society - the nature of the social relationships between people who lived in the towns?

The consideration of these questions consti-
tutes potentially one of the most fruitful lines of investigation open to both urban geographers and social historians and is, by definition, full of interdisciplinary implications and problems. On the one hand, those historians who have so far attempted to grapple with this issue have suggested various

explanations which Michael Anderson (1976, p.334, note 59) - in one of his many appeals for a more sociologically-rigorous form of social history - had no difficulty in describing as 'equally plausible but mutually contradictory'. Reciprocally, the most recent work in the field by urban geographers has either considered the problem in a twentieth-century context (is the planner the servant or master of social processes?), or has tried to relate analysis of census data to the problem of where, precisely, should nineteenth-century towns be put on the continuum from pre-industrial to post-industrial cities, (Harvey, 1975 pp. 44-5; Shaw, 1977, p.335; Carter and Wheatley, 1978, pp.51-3).   In spite of these interests, explicit discussion of the links between the shapes on the ground and the shapes in society remains at a relatively primitive stage of development.   This chapter considers four aspects of the problem.   First, it discusses - from the standpoint of the historian - some of the difficulties which arise in any attempt to establish the nature of the shapes on the ground. Secondly it examines the analagous problem of establishing the identity of the shapes in society.   Thirdly, the attempts which historians have so far made to link the two are described.   Finally, some general comments and suggestions are offered on the basis of this analysis.

Shapes on the ground

At one level of inquiry, historians might choose to ignore entirely the question of what the shapes on the ground in nineteenth-century cities were actually like.   They might begin, instead, by accepting the consensus of mid-Victorian contemporaries that their new 'great cities' were indeed segregated to an unprecedented extent.   After all, the evidence for that conclusion is considerable. The practice of equating working-class areas with 'darkest Africa' and the fear of the unknown and irreligious masses who inhabited these regions bear witness to the perception, by middle-class contemporaries, of an unprecedentedly segregated, urban society which had evolved by the middle of the nineteenth century (Briggs, 1968, pp.59-64). McCulloch, for instance, writing of Sheffield, noted the extreme contrast between the 'dingy, mean appearance' of the town as a whole, and the 'extreme beauty of the surrounding country embellished as it is in every direction by the numerous villas of the

236

opulent bankers, merchants and manufacturers of Sheffield', a description almost identical to Engels' more famous contemporary piece on Manchester. James Smith's account of the sanitary state of Leeds in 1845 painted a very similar picture (Cannadine, 1977, pp.460-1).

If a historian's prime concern was, for example, to understand the 'Condition of England' question of the 1840s, he might argue that the 'objective' facts of the situation were largely irrelevant, and that what most concerned him was what contemporaries *thought* was going on, or as David Harvey (1975, p.33) has described it, 'the cognitive state of the individual with respect to his spatial environment'. Whether their perception of their circumstances was empirically valid or not would be largely irrelevant - if the object of the exercise was to establish what contemporaries thought and what, on the basis of their thoughts, they then did. So, on the basis of their belief that cities were unprecedentedly segregated, the historian might then go on to explain the 'Condition of England' question largely in terms of the dis-covery by the rich of the existence of the poor, from whose houses and residential areas they felt themselves to be segregated. And so he could happily accept their belief that mid-Victorian cities were more highly differentiated than ever before, without ever really needing to probe beyond their image of the city to the reality.

But supposing the historian argued that what was important was not what contemporaries thought was happening, but the objective reality of the situation, which may in fact have influenced them more - albeit unconsciously - than their own deliberate decisions made on the basis of their own conscious perception. Under these circumstances, he might turn to his colleagues in historical geography for elucidiation. In particular, he might look at the work of Richard Lawton, (1955, p.110), Colin Pooley (1977, p.364) and Harold Carter (1978)on mid-Victorian Liverpool and Merthyr, and he would be encouraged to note their conclusion that 'the rapidly-growing city of the mid-Victorian period exhibited a high degree of residential differentia-tion: the main social dimensions of city structure had a clear spatial expression, and were reflected in distinctive social areas' (Pooley, 1977, p.364). Although the historian would appreciate the geographers' plea for caution because of the problems of the census data and the difficulty in agreeing on a suitable unit of spatial analysis, he could still happily conclude, on the basis of their analysis, that the mid-Victorian observers had basically got things right, and that what they *thought* was happening was closely consistent with what was *actually* happening.

But if he turned to the work of another historical geographer, David Ward, he would immediately be thrown into some confusion. For Ward (1975, pp. 137-8, 142, 1976, p.331) has argued that mid-Victorian cities in fact displayed a sur- prisingly low degree of residential differentiation; that single suburbs contained a great range of in- habitants as judged by the criteria of income, status or occupation; and that in general, the social geography of towns was very 'weakly defined'. Nor is this an argument based on theoretical rather than empirical evidence, for his analysis of the census data in Leeds for the period from 1841 to 1871, leads him to conclusions - on the basis of similar evidence - strikingly different from those of his colleagues:

> About half of those inhabitants with middle-class occupations live in relatively well-defined quarters; the remainder are inter- spersed with the less affluent. This less affluent majority was certainly not as well sorted by gradations of skill or any other measure of strata differences as was suggested by descriptions of Leeds at the turn of the century (Ward, 1977).

So, if he is correct, contemporaries had got it wrong: their perception of their circumstances did not square with the reality. Mid nineteenth- century English towns were not like the Chicago of Burgess *et al*, and the interesting question for the historian then becomes: how was it that contem- poraries came to be so misled?

This divergence of opinion among historical geographers, well illustrates the common problem of the historian when he wheels his trolley around the supermarket of ideas in search of theoretical packages or bodies of data from other disciplines which may help him order his own inchoate mass of evidence: namely that in most other disciplines, from whose findings and theories he may wish to borrow, there is no agreement as to just what those theories and findings are. So, while one part of the geographical profession might applaud him if he decided to accept the Lawton-Pooley-Shaw-Carter argument and use it as a basis on which to make

238

various observations and explanations about the causes and consequences of class conflict in mid nineteenth-century England, another part might feel that the whole enterprise was founded on mistaken premises.

This, then, is one difficulty facing the historian approaching the work of historical geographers: namely, whom should he believe? But a second difficulty, which arises regardless of which conclusions he accepts, is the question of whether it is actually possible - or valuable - to try to establish the 'objective' characteristics of residential areas on the basis of census data. Let me give two examples. If one took Mayfair and Edgbaston, these might both be cited as ideal instances of highly-segregated, upper- and upper middle-class urban areas, whose characteristics and reputation were already well-established by the middle years of the nineteenth century. Mayfair, after all, was reputed to be the most valuable piece of real estate in the world, and was popularly thought to embody all that was wealthy, opulent, splendid and exclusive in London high society (Colby, 1966). In the same way, Edgbaston, the 'Belgravia of Birmingham', was the home of the town's nonconformist elite, and had no real rival for most of the nineteenth century as a bastion of suburban exclusiveness and middle-class values (Cannadine, 1977, pp.471-2). *Subjectively,* both of these communities were highly segregated from the rest of the urban area. Indeed, it was in part the perceptions by contemporaries who lived in places such as these which caused them to hold forth so eloquently on the subject of the 'two nations'.

But *objectively,* the characteristics of these areas were rather different. Here, for example, is a recent analysis of Mayfair on the basis of the 1841 census:

> Only a very small proportion of the residents on the estate (some five to ten per cent) belonged to the titled of leisured classes. The social *cachet* of a good address there might still, in Victorian times, be as highly prized as ever; but even in this citadel of the *beau monde* such residents were far outnumbered by the rest of the population (Sheppard, 1977, pp.93-5).

Indeed, in 1871, some sixty-three per cent of the households in Mayfair had no domestic servants: the actual part of the estate which 'objectively'

conformed to its popular image was but a very small
part of the whole.   In fact the area immediately to
the south of Oxford Street had more in common with
working-class suburbs than it did with the upper-
class part of Mayfair itself.   In the same way, at
Edgbaston, half of the houses constructed on the
Calthorpe estate were of a decidely lower middle-
class or artisan type, far distant socially - but
not spatially - from the dwellings of the nonconformist elite (Cannadine, 1977, p.474).   Like
Mayfair, the social spectrum of the Calthorpe estate
was far broader than its overall 'tone' suggested.
For instance, the itinerant life of the young
J.R.R. Tolkien in the early 1900s, on the borders
not only of Edgbaston but also of respectability,
was a world away in style - but not in location -
from the elegant, leisured life of a man like
Joseph Gillott, with his collection of paintings by
Turner and his lavish wine cellar (Carpenter, 1978,
ch. 2).   As with Mayfair, some parts of Edgbaston
had more in common with the neighbouring lower-
class areas than they did with that central part of
the suburb from which its high popular reputation
was derived.
        What are we to make of this?   One answer might
be to stress that there are at least two concepts
of residential differentiation valid for the nineteenth as for the twentieth-century city.   The
first might be labelled 'objective', concerned with
criteria such as income, occupation, status,
ethnicity and the like.   The second is 'subjective',
stressing the mental maps drawn by contemporaries -
the personal space of the actors, rather than the
statistical space of the census (Gould and White,
1974, ch. 1).   Moreover, there would be many more
mental maps than patterns revealed by analysing the
census.   And it might be that under these circumstances, the best conclusion to reach is that
'subjectively' the shapes on the ground stood out
more vividly to contemporaries than 'objectively'
they can be revealed to us today.   Perhaps, too,
it is with the dichotomy between these two
evaluations of the same area that historians in
particular should be concerned, asking how it was
that, despite their relatively broad social spectrum,
certain areas were able to establish and retain a
'subjective' reputation for exclusiveness which the
'objective' analysis of their inhabitants in fact
belied.

Shapes in society

If the historian of the nineteenth-century city is
in debt to the historical geographer for much of
what he knows - and much of what confuses him -
about the spatial structure of towns, he is even
more bewilderingly indebted to the sociologist for
assistance in delineating the shapes in society.
For the sociology section in the supermarket of
ideas is even more crammed with conflicting products
than are the shelves on which the historical
geographers display their wares.   In the first
place, there is the problem of the way in which the
term 'class' should be employed.   It can be used
simply as a descriptive term, denoting 'the
statistical category of people united by having
certain objective criteria of stratification, not
conscious of itself as such, and not doing anything
as such' (Vincent, 1967, p.21).   Or one might
argue that the more important definition is class
as class consciousness; that actors must be aware
of their class identity and act upon that basis
(Thompson, 1978, p.149).   The first of these
definitions, which usually implies analysis of
census data, clearly has some affinity with the
'objective' areas of towns which the historical
geographers, using the same data, seek to identify.
The second, which assumes that perception is the
pre-condition of action, clearly has more in common
with the 'subjective' analysis of neighbourhood.
    Most social historians who use the concept of
class would accept that the only valid definition
to take is that of class as class consciousness, as
a concept which does not so much describe a person's
objective circumstances, but his perception of his
situation, and the action which he takes on the
basis of that evaluation.   But that then leaves
the problem of the definition of class.   One such
definition used by historians derives from the
original Marxist one, which argues that 'the class
experience is largely determined by the productive
relations into which men are born - or enter
voluntarily' (Thompson, 1968, p.10).   Thompson
himself, John Foster (1968, pp.282, 291) and Harold
Perkin (1969, p.176), who describes class as
'vertical antagonism between a small number of
horizontal groups each based on a common source of
income', might all be said to use class in this
sense.   But another historical tradition, more
indebted to Max Weber and exemplified in the work of
Dahrendorf (1959, pp.172-4) and Neale (1972, p.9),

stresses the difference between social stratification, in large part influenced by relation to the means of production, and class, which they see as deriving from conflict over power and authority within 'imperatively co-ordinated associations'. In short, they reject the Marxist assumption that there is an implicit and direct causal link between economic circumstance and class consciousness.

Moreover, both of these definitional problems ignore the more general question of whether one should seek to see the shapes in society in terms of class analysis at all. Marxist historians would argue that one should; that society is founded on conflict; and that out of conflict was born class consciousness (Thompson, 1978, p.149). For them class consciousness and class conflict represent the fundamental state of society, and the crucial historical enterprise must be to ask how and when it came into being. If it cannot be found, then it is its absence which must be explained, by such developments as *embourgeousement* or a labour aristocracy (Moorhouse, 1978; Crossick, 1976). It is important to notice here that a concept of class which sees it, historically, as 'observation of the social process over time', must cast doubt on the possibility of utilizing census data at all for this purpose (Thompson, 1978, p.147). For whatever shapes on the ground that data may reveal, it does so in an essentially static manner, which by definition must be hard to link to a dynamic concept of shapes in society.

But while this historical view of the shapes in society derives from a school of sociology which sees conflict as central to the nature of social relationships, the alternative sociological tradition, of the structural-functionalists, has also found its historical adherents. For them, class can at most be nothing more than an analytical, descriptive category, explaining nothing about the functioning of society at all. The social system, according to this analysis, is normally not in conflict but in equilibrium, largely because there are no simple, clear-cut distinctions between a few classes, but a series of layers or strata without clearly marked boundaries or divisions (Parsons and Shils, 1954, pp.209-33). Among historians who have adhered to this analysis of social relations, one might mention Currie and Hartwell (1965, pp.637-9) in their review of E.P. Thompson, Patrick Joyce's (1975) picture of social relations in nineteenth-century Blackburn, and D.S. Gadian's (1978) and A.E. Musson's (1976, p.336) recent critiques of

John Foster in which, arguing against the existence of class and class consciousness as the fundamental relation in society, they describe 'a multiplicity of classes merging into each other, without clear-cut divisions, ... and the better off artisans were nearer in economic and social status to small masters, shopkeepers, etc., than to the lower labouring classes'.

The third approach which historians have taken, influenced by sociologists like Dahrendorf (1959, pp.159, 163) and Lockwood, is to argue that the consensus and conflict models are not mutually exclusive - as some of their most committed supporters would claim - but that they are merely the opposite sides of the same coin. This approach argues that to set up a model which assumes that class, or no class, is the fundamental, 'true' form of social relationship misleads because of its very claim to comprehensiveness. Among social historians, both Professors Perkin and Best, in their accounts of mid-Victorian England, have inclined to this view. Professor Perkin (1969, ch.9), although arguing that a class society was very definitely in being by this time, suggests that it was only 'viable' (i.e. held together) because the ties which bound the classes together and promoted consensus were at least as strong as the forces which drove them apart in conflict. In the same way, Professor Best (1971 p.xvi) concluded that while 'feelings of class antagonism were obviously there - feelings largely economic in source, and representing, so to speak, "horizontal" social divisions', they were 'not often, or for many people, stronger feelings than those attaching to their, so to speak, "vertical" connections with classes below and above them'.

Historical studies of the links between society and urban form

Accordingly, if the geographers are unable to agree on the shapes on the ground and the sociologists still quarrel energetically about the shapes in society, what is the historian - vainly trying to link them - to do? In practice, historians have not been deterred by the evident methodological problems, and have happily attempted to relate the physical and social geometry of nineteenth-century towns. Most have begun by assuming that a particular spatial form is 'given' and that it then exerts influence on social relationships (Harvey, 1975, pp.27, 44; Gregory, 1978, p.86). Indeed, on the basis of this, four different interpretations are possible. Two of these, indebted to the Lawton-Pooley-Shaw-Carter school of geographers - are concerned with the influence on social behaviour of a relatively high degree of residential segregation. The other pair - consciously or unconsciously echoing Ward - dwell on the social consequences of there being little or no residential segregation.

One argument, advanced by Foster, centres on the assumption that the necessary pre-conditon for the rise of class consciousness was a real and immediate perception of deprivation. In order to feel poor or deprived, so this argument runs, it is necessary to be physically close to wealth and privilege (Foster, 1968, pp.291, 293; Runciman, 1966, pp.11, 16, 19). The comparative reference group, in this model, is the 'resident bourgeoisie'. And it is their close proximity, rather than any shared feelings of solidarity between individuals of the same socio-economic status, which is the chief motive force behind the growth of class consciousness. It is because the poor man knows he is at the gates of the rich man, rather than because he is aware that there are many other poor men like him, that his sense of grievance is aroused. So, on the basis of this argument, it is a reasonable assumption that towns which have the most undifferentiated residential patterns might be expected to witness the greatest degree of class conflict.

But the same evidence has been used to evolve a completely contradictory argument. This suggests that if rich and poor, bosses and workforce, are all residentially mixed up, some form of face-to-face community is perpetuated, albeit in an urban setting, which offers the possibility of real co-operation and collaboration in the manner which the structural-functionalists would recognize. This spatial arrangement avoids the bleak isolation which may result from only those of the same class being located together residentially. As Patrick Joyce (1975,pp.546-8) has shown, in an environment thus spatially constituted, there was the very real possibility of the successful exercise of paternalism in what was, in effect, an industrial village within the context of a larger urban area. In this model, then, it is the perception, not of relative deprivation, but of relative proximity to, and close involvement with, those of higher and lower social status, which actually served to retard or prevent the growth of class consciousness. By this analysis, a town in which all the classes were mixed up - as Briggs (1957, pp.292-302) implies was the case in early nineteenth-century Birmingham - is a much more likely scene of class collaboration than of class conflict.

On the other hand, if one assumes initially a high degree of residential differentiation, the same contradictory conclusions may also be reached. If the initial assumption is that mixing up people of different socio-economic status prevents the evolution of class consciousness, then the corollary is that it is by separating out the classes spatially that they will come to perceive their true social identity. Rather than a strong sense of relative deprivation, it is the lack of the proximate civilizing influence of the middle class combined with an awareness of solidarity with one's peer group which is central. As Professor Perkin (1969 p.174) has put it: 'the geographical separation of the classes, and the mutual ignorance, suspicion and musunderstanding which went with it, were a powerful factor in the rise of class conflict in the early-nineteenth century'. In other words, the more segregated the city, the greater the likelihood of class consciousness and conflict.

But if one assumes that it is by mixing up the classes that class consciousness and conflict are most likely to be engendered by an awareness of relative deprivation and social injustice, then it follows that if the classes are kept apart, class conflict will be minimized. By this argument it is the crossing of spatial boundaries which heightens social tension. For example, nineteenth-century London was the most highly-segregated city in England, yet was one with arguably the weakest class consciousness. Stedman-Jones (1976, pp.247, 346) explains this in terms of its undeveloped economic sub-structure. But, at least theoretically, it might equally be explained by its strongly-developed patterns of residential segregation. For if you live in an area where everyone is of similar socio-economic status, it becomes much harder to devise a reference group by comparison with which you may feel deprived.

Some general comments

These four models may be most easily summarized by drawing up a two-by-two table as follows:

	Class Conflict	Class Collaboration
Segregation	X	X
No segregation	X	X

The possible options open are these: following the
top line of the table, one might argue that the
existence of shapes on the ground did - or did not -
lead to shapes in society; and following the second
line, one might argue that the lack of shapes on the
ground did - or did not - lead to shapes in society.

Of course, other options of a more complex kind
are open within the limits of these variables. Let
us continue to assume that it is the shapes on the
ground - however measured, and whether they existed
in reality or not - which fundamentally influenced
the shapes in society. But let us now go on to
incorporate in the model the fact that the patterns
on the ground did themselves change. Empirically,
in the nineteenth century, they did so in only one
direction: from none- or very little - segregation
to some - or much greater - segregation. Lawton
and Ward would both accept that: the difficulty
centres on when the most significant shift took
place. On the basis of this change in spatial
patterns, and using the causal connections just
discussed, two possibilities are open. One can
argue, first,that the move from non-segregation,
which produced class conflict, to segregation, in
which there was collaboration, is of major - and
unexplored - importance in explaining the shift from
the class conflict of the 1830s and 1840s to the
calmer mid-Victorian period. To make this argument
would oblige one to accept the Lawton model as far
as the chronology of segregation was concerned. But
if one accepted Ward's argument, one could suggest
that it was the shift from a little-segregated mid-
Victorian period, in which there was in consequence
no real class conflict, to the highly segregated
late-Victorian period, where there was severe class
conflict, which was the crucial change. Theoretically - and indeed empirically on the basis of
the conflicting data which the historical geographers have produced so far - both of these explanations merit equal consideration.

But a second response to these arguments is not
to make them increasingly complex, but to stand
them all on their heads. It is equally plausible
to assume that 'spatial structures *realize* social
structures'; that it is the social process which is
the principal determinant of spatial form, and not

246

the other way round (Harvey, 1975, pp. 44-5;
Gregory, 1978, p.119).   Under those circumstances,
the two-by-two table comes out as follows:-

	Segregation	No Segregation
Class Conflict	X	X
Class Collaboration	X	X

It would be tedious to go through the causal
options open here in detail: but here is a bald
summary.

The most obvious argument is that an individual,
already aware of his increased social distance from
his former peer group - as a result of increased
income, perhaps - adds a spatial dimension to his
social separation by leaving the district and moving
to a higher-class area (Crossick, 1976, p.313).   In
this case, the identity aspirations come first, as
a result of changed economic circumstances, and are
then expressed spatially (Timms, 1971, p.251).
Class awareness or consciousness is here a powerful
motor of social exclusiveness.   But one might also
argue that an actor whose material circumstances
have greatly improved would wish to get as close as
possible to the 'leaders of society', to segregate
himself from those he felt to be 'beneath' him and
to live nearer his 'betters' (Hoyt, 1959, p.501).
Under these circumstances, one would not stress the
exclusive nature of a social climber's destination,
but the width of social spectrum there.   Seen from
this standpoint, for instance, Edgbaston was a
broad social spectrum, in which Tolkien and others
could get close to the leaders of society.   A third
argument would be that while economically there was
great class collaboration, in terms of status men
still sought to mark themselves off from their
colleagues; that the very fact of shared economic
interest was itself likely to increase the induce-
ment to show separateness in other ways.   Finally -
reversing the lines of causality in the Patrick
Joyce model - one might argue that identity of
economic interests did tend to encourage masters and
men to live together.

Conclusion

The point of this contribution has been merely to
show that, on the basis of the research at present
available, we have no coherent body of theory concerning the links between spatial segregation and
social class.   Different hypotheses have indeed

247

been advanced on the basis of specific pieces of detailed research, but they are so contradictory, and indeed mutually exclusive, that they cannot be said, collectively, to add up to anything. If one of the purposes of social history is both to test and to generate theory, then as far as this particular subject is concerned, we have only reached the stage of recognizing that none of the theories and hypotheses yet propounded is satisfactory.

Where, then, do we go from here if we wish to generate some comprehensive theory on this subject? Four general suggestions may serve as a speculative conclusion. The first is to consider the possibility that there is really no connection at all between spatial and social geometry; that there are no causal links between them running in any direction; that there is no theory to be found. One might then accept Professor Best's (1971, pp.15, 17) agnostic view that there was class conflict and also class collaboration, and that cities were segregated as well as having their inhabitants from varied backgrounds and of differing status crammed together. It is, after all, important to remember that the assumption of a causal connection - in either direction - is easily made; but that the theory and evidence to validate such a model are still extremely sparse. In practice, we do not really know what - if anything - is intrinsic to the urbanization process which links the shapes on the ground with those in society.

A second possibility is to assume that, while there might be links between these two types of shapes, they are relatively unimportant. One might argue that the productive relationship at the work place is of far greater significance in bringing class consciousness about than any system of residential spatial patterns could ever be; or that matters such as topography were far more significant in determining the spatial pattern of cities than the working out of class consciousness in physical terms. Again the implication is that searching for a theory to link spatial and social geometry is not a very important enterprise.

A third option is to accept the (unproven) claims of Marxist geographers and the *Urban History Yearbook* that there must indeed be a connection between place and process, but that the likelihood is that the lines of causality do not all run one way. Therefore we should see the links between them as being mutually reinforcing: 'Spatial structure is not...merely the arena within which

248

class conflicts express themselves, but also the domain within which - and in part through which - class relations are constituted' (Gregory, 1978, p.120).

Even if we accept that there may be other influences on spatial as on social patterns, and that the links between the two - insofar as there are any - are complex, that only enables us to state the central problem: Given that the shapes on the ground and in society are so hard to describe, and given that the links between them are both complicated and limited, then how in practice do we devise a research project to investigate them?

References

Anderson, M. (1976) 'Sociological history and the working-class family: Smelser re-visited', *Social History,* 1, 317-34

Best, G.F.A. (1971) *Mid-Victorian Britain, 1851-1875,* Weidenfeld and Nicolson, London

Briggs, A. (1957) 'The background of the Parliamentary Reform Movement in three English cities (1830-2)' *Cambridge Historical Journal,* 18, 292-317

Briggs, A. (1968) *Victorian Cities,* Penguin, Harmondsworth

Cannadine, D. (1977) 'Victorian cities: how different?', *Social History,* 2, 457-82

Carpenter, H. (1978) *J.R.R. Tolkien: a biography,* Unwin, London

Carter, H. and Wheatley, S. (1978) 'Merthr Tydfil in 1851: a study of spatial structure', *Social and Residential Areas of Merthr Tydfil: Working Paper* 2, U.C.W. Aberystwyth

Colby, R. (1966) *Mayfair: a town within London,* Country Life, London

Crossick, G. (1976) 'The Labour Aristocracy and its values: a study of mid-Victorian Kentish London', *Victorian Studies,* 19, 301-28

Currie, R. and Hartwell, R.M. (1965) 'The making of the English working class?', *Economic History Review,* second series, 18, 633-43

Dahrendorf, R. (1959) *Class and class conflict in industrial society,* Stanford University Press, Stanford, Calif.

Donnelly, F.K. (1976) 'Ideology and early English working-class history: Edward Thompson and his critics', *Social History,* 1, 219-38

Dyos, H.J. (ed.) (1978) *Urban History Yearbook,* Leicester University Press, Leicester

Foster, J. (1968) 'Nineteenth-century towns: a class dimension', in H.J. Dyos (ed.), *The Study of Urban History,* Arnold, London, pp.281-291

Gadian, D.S. (1978) 'Class consciousness in Oldham and other North-West industrial towns, 1830-1850', *Historical*

Journal, 21, 161-72

Giddens, A. (1973) *The class structure of advanced societies,* Cambridge University Press, Cambridge

Gould, P. and White, R. (1974) *Mental Maps,* Penguin, Harmondsworth

Gregory, D. (1978) *Ideology, science and human geography,* Hutchinson, London

Harvey, D. (1975) *Social Justice and the City,* Arnold, London

Hoyt, H. (1959) 'The pattern of movement of residential rental neighbourhoods', H.M. Mayer and C.F. Kohn, (eds.) *Readings in urban geography,* Chicago University Press, Chicago, Ill.

Jones, G.S. (1976) *Outcast London: a study in the relationship between classes in Victorian society,* Penguin, Harmondsworth

Joyce, P. (1975) 'The factory politics of Lancashire in the later nineteenth century', *Historical Journal,* 18, 525-56

Lawton, R. (1955) 'The population of Liverpool in mid-nineteenth century', *Transactions of the Historic Society of Lancashire and Cheshire* 107, 89-120

Moorhouse, H.F. (1978) 'The Marxist theory of the Labour Aristocracy', *Social History,* 3, 61-82

Musson, A.E. (1976) 'Class struggle and the Labour Aristocracy', *Social History,* 1, 335-56

Neale, R.S. (1972) *Class and ideology in the nineteenth century,* Routledge and Kegan Paul, London

Parsons, T. and Shils, E.A. (1951) *Toward a general theory of action,* Harvard University Press, Cambridge, Mass.

Perkin, H. (1969) *The origins of modern English society, 1780-1880,* Routledge and Kegan Paul, London

Pooley, C.G. (1977) 'The residential segregation of migrant communities in mid-Victorian Liverpool', *Transactions of the Institute of British Geographers,* N.S. 2, 364-82

Runciman, W.G. (1966) *Relative deprivation and social justice,* Routledge and Kegan Paul, London

Thompson, E.P. (1968) *The making of the English working class,* Penguin, Harmondsworth

Thompson, E.P. (1978) 'Eighteenth-century English society: class struggle without class', *Social History,* 3, 133-65

Shaw, M. (1977) 'The Ecology of social change: Wolverhampton, 1851-71', *Transactions of the Institute of British Geographers,* N.S. 2, 332-48

Sheppard, F.H.W. (ed.) (1977) *The survey of London, vol. 39. The Grosvenor Estate in Mayfair. Part 1, General Survey,* University of London, Athlone Press, London

Timms, D.W.G. (1971) *The urban mosaic: towards a theory of residential differentiation,* Cambridge University Press, Cambridge

Vincent J.R. (1967) *Pollbooks: how Victorians voted,* Cambridge University Press, Cambridge

Ward, D. (1975) 'Victorian cities: how modern?', *Journal of Historical Geography*, 1, 135-51

Ward, D. (1976) 'The Victorian slum: an enduring myth?'. *Annals of the Association of American Geographers*, 66, 323-36

Ward, D. (1977) Letter to the author, 16 May, 1977

Chapter 10

STABILITY AND CHANGE IN URBAN COMMUNITIES: A GEOGRAPHICAL PERSPECTIVE

Richard Dennis

Introduction

Most of us find 'change' a more attractive object of
study than 'stability': thus many historical
geographers have regarded Victorian cities as
'transitional' between pre-industrial and industrial
stereotypes and, however short their period of study,
have frantically searched for the slightest change
that would confirm their theories about the levels
of residential segregation that were expected to
accompany industrialization, specialization and
alterations in social structure.   Several recent
pieces of work have taken this approach.   Warnes'
(1973) treatment of Chorley and Pritchard's (1976)
study of Leicester both cover sufficiently long or
critical periods for our expectations of change to
be justifiable.   Other work, such as Mark Shaw's
ecology of social change in Wolverhampton (Shaw,
1977) and some of my own research on Huddersfield
(Dennis, 1976), reflects dependence on the manuscript census by searching almost too diligently for
change during an all too brief twenty years in the
middle of the century.   A third category of
research includes cross-sectional studies at a
single date, usually 1851 or 1871 (Carter and
Wheatley, 1978a, 1978b; Lawton and Pooley, 1975b);
yet here too the results are customarily interpreted
on the assumption that the cities to which they
refer were 'in transition' from a Sjobergian 'pre-
industrial' city, in which the rich dominated the
centre and the poor made do with the periphery, to
a 'modern' city in which the locations of rich and
poor were reversed.   The only difference between
these studies lies in the distance along the transition at which it is concluded that each city can be
located.
    The first half of this chapter argues that some

of the hypotheses of transition, adopted in particular by geographers, have obscured our understanding of socio-spatial change more than they have enlightened it. It is also suggested that the period on which most research has been concentrated is probably the least interesting with regard to change, and that we have been too concerned with quantitative methodology at the expense of studying the mechanisms of change.

The reversal of spatial structure

At the ecological scale geographers have concentrated on charting the changing nature of social areas in terms of one or more of the following ideas:-

(1) a reversal of spatial patterns of socio-economic status;
(2) an increase in the scale of residential segregation; and
(3) a separation of axes of socio-economic, family and ethnic status, corresponding to Timms' (1971), McElrath's (1968) and Abu Lughod's (1969) elaborations of social area theory (Shevky and Bell, 1955).

The reversal of spatial structure

The first of these three ideas is derived from two simple stereotypes of 'pre-industrial' and 'modern' cities advanced by Sjoberg (1960) and Burgess (1925). It is argued that as population, and therefore the demand for housing, increased and as transport technology improved, so there was a reversal of the location of high- and low-status areas. But, if the dimensions of urban spatial structure were changing in accordance with the third idea listed above, and if social structure was changing in response to industrialization, increased occupational specialization, the growth of bureaucracy and the politicization of the labouring classes, then it is unlikely that the same collection of variables is equally appropriate for the definition of 'class' or 'status' at different periods. I do not propose to embark on yet another critique of the use of the Registrar General's 1951 scheme, as advocated by Armstrong (see Armstrong, 1966, 1972; Royle, 1977; Holmes and Armstrong, 1978); rather I am arguing that whatever definition of status we adopt for, say, 1851, then it will be less appropriate for the same town in 1801 or 1901.

254

The advantage of keeping the definition unchanged
is, of course, that it facilitates quantitative
comparison, but if it emerges that we are really
comparing unlike quantities because contemporaries
did not define status as we do or changed the
criteria they used for ranking their fellows, then
the advantage seems to be a pretty hollow one.

A further criticism of the status-reversal
hypothesis is that it distracts our attention from a
much more fundamental change which took place in the
form of any city during the nineteenth century.
Whatever the changes in the city of 1800 that
became the inner city of 1900, its physical
expansion between those dates was far more important.
It does not follow that because the city got bigger,
the scale of segregation inevitably increased: both
Vance (1966, 1977) and Ward (1975) have noted the
growth of cities by 'cellular reproduction'. But
it does follow that the so called reversal of the
status gradient involved little more than the displacement of the rich by the growth of a non-
residential Central Business District, rather than a
displacement of the rich by the poor. The poor
remained where they had always lived, on the edge of
the old city which became the innermost residential
area of the new. Subsequently they may have moved
out to displace the middle classes as the outer
suburbs of the mid-nineteenth century became the
inner city of today, but that process was one of
wave-like invasion and succession, not a reversal of
the spatial structure of the city. The implication
of this observation is that we have devoted too
much attention to the purely residential aspects of
Schnore's (1965) and Johnston's (1972) models of
spatial change, and too little to the relationships
between the growth of central commercial functions
and the exodus of the middle classes.

It is strange that so much research by
geographers continues to assume Sjoberg's and
Burgess' models as the end-points of change, when
the relevance of each to actual pre-industrial or
twentieth-century British cities has been widely
questioned (Langton, 1975; Robson, 1972, 1975). Few
British towns, certainly by the eighteenth century,
had centrally-located, non-productive elites, and
Vance's cellular model of a capitalist, but still
pre-industrial city would appear a more promising
starting-point (Vance, 1971). The evidence of
social area analyses and factorial ecology suggests
that socio-economic status in the modern city is
distributed sectorally, more in the manner of Hoyt's
'high rent' areas, rather than concentrically as in

255

Burgess's model (Robson, 1973, 1975). In fact, if
we hypothesize a transition from Vance's cellular
city to Harris and Ullman's multi-nucleated modern
city (Harris and Ullman, 1945), there is little
reorganization of spatial structure to be explained,
except within each cell or around each nucleus.

The scale of segregation

It is clear that the spatial segregation of status
groups did not begin suddenly in 1800, but the awareness of segregation, the attitudes towards it and
the scale at which it occurred did all evolve during
the century (Best, 1971; Dennis, 1980; Harrison,
1971; Ward, 1975, 1976, 1978, 1980). Early-nineteenth century forms of social segregation,
vertically within dwellings and horizontally between
front-street and court dwellings, had, by mid-
century, evolved into segregation by streets or
estates, although these effects were still often too
detailed to show up clearly in analyses at the scale
of enumeration districts. Only in the largest
cities, such as Manchester, or at the extremes of
status - middle-class suburb and immigrant slum -
were homogeneous social areas of the type expected
by contemporary urban theory to be found.
     Several geographers and social historians have
attempted to measure the relative degree of segregation of different social groups in the city but, as
yet, there has been little attempt to trace changes
in levels of segregation through time. Foster (1974)
has noted that in 1841, 37 Irish families in Oldham
had Irish neighbours but 152 did not. By 1861,
576 Irish families lived next door to other Irish
and only 372 had English neighbours. The implication is that, as this most detailed level, the Irish
were becoming more segregated. Rather more
sophisticated statistical analysis has been applied
by Carter and Wheatley (1980) to Merthyr Tydfil in
1851, by Pooley (1977) to the segregation of migrant
groups in 1871 Liverpool, and by Daunton (1977) to
occupational groups and 'classes' in Cardiff in the
same year, while Ward (1980) has examined class
segregation at several different scales in Leeds
between 1841 and 1871.
     Although the absolute sizes of the ethnic or
social groups being compared do not affect the
values of Dissimilarity or Segregation Indices
(Peach, 1975), they influence the interpretation
that can be placed upon them. If, for instance,
two studies at the same scale each revealed an index
of 50 for the Irish-born, (i.e. in each case, fifty

256

per cent of Irish would need to move areas for their distribution to parallel that of the rest of the population) we would make very different interpretations if in one case we were dealing with 50 Irish and in the other with 5000.

The measurement of segregation, like that of status-reversal, is complicated because variables that are defined in the same way at different dates may have changed in their meaning. For example, in 1841 or 1851, when most Irish in English towns were first-generation migrants, the variable 'Irish-born' is a reliable indicator of Irish culture. But by 1871, many who thought of themselves as Irish had in fact been born in England of Irish parentage and the 'Irish born' variable may have included a smaller proportion of those subscribing to a distinctively 'Irish' culture. This would confuse calculations of the changing degree of segregation of the Irish through time, and it would prove an almost impossible task of record linkage to identify second-generation immigrants by tracing them as children in their parental home in an earlier census.

One possible method for studying changes in the scale of segregation would involve the calculation of segregation indices at a series of dates and at a number of scales. The hypothesis that distinctive social areas increased in size with the passage of time could be tested by noting the periods of greatest change in index values at different scales. We would always expect values to be higher for analysis at the scale of the street than for enumeration districts, and higher for enumeration districts than for wards (Poole and Boal, 1973), but we might find periods in which index values at one scale changed more rapidly than values obtained at other scales.

Social area theory

The third statement - about the relationships between different aspects of social differentiation - is perhaps the most interesting, and has received considerable attention in studies of Liverpool (Lawton and Pooley, 1975b, 1976), Merthyr Tydfil (Carter and Wheatley, 1978a, 1978b), Chorley (Warnes, 1973) and Wolverhampton (Shaw, 1977). The argument put forward by Timms in *The Urban Mosaic* (1971, p.146) was that in 'feudal cities' there was only one axis of residential differentiation, and variables relating to social rank, family status, ethnicity and migration status were intercorrelated, whereas in 'modern cities' each of these sets of variables constitutes an independent axis of differentiation. The argument can be extended by considering the spatial patterns formed by each set of variables. The 'feudal city' was characterized by a simple contrast between centre and periphery, while in the 'modern city' social rank is distributed sectorally with clearly defined high- and low-status sectors, family status is distributed concentrically, the basic distinction being between a 'disorganized' core of young single people, the widowed, childless couples, etc. and a family-oriented periphery, and migrant and ethnic minorities are found in randomly located clusters. Between these two extremes lie other types of city, labelled by Timms as 'colonial', 'immigrant', 'pre-industrial' and 'industrializing', and we may presume that nineteenth-century cities belong somewhere in these groups. For example, Carter and Wheatley (1978b) have made the case for Merthyr as a 'colonial city', using the notion of 'internal colonialism'.

These hypotheses are most frequently tested by means of factor analysis, but while we may apply this technique satisfactorily to a single data set relating to one point in time, problems arise when we attempt to compare successive censuses in order to see whether there had been any appreciable development in the direction of the 'modern' city. Shaw's study of Wolverhampton in 1851, 1861 and 1871, and my own more limited comparison of Huddersfield in 1851 and 1861 (Dennis, 1975, 1976), used identical data sets for each census, but then encountered the problem of comparing similar, but not identical sets of factor loadings for each year. While the relationship between factors can be des- cribed using 'congruency coefficients' (a variant of correlation coefficients, measuring the correla- tion between, say, factor I in 1851 and factor I in 1861), it is difficult to reach any firm conclusions about changes in spatial structure simply by compar- ing the distributions of factor scores for two or more years.

A more fundamental problem is that the available manuscript censuses cover too short a period for analyses based on their comparison to do more than hint at structural or spatial changes (Dennis, 1979); but if we extend our comparison beyond the 1851-71 period, data problems restrict the inferences that we can make about longer-term changes. Warnes (1973), for example, compared the 1851 census for Chorley with an 1816 Vestry Committee Survey, but the 1816 survey had a more limited range of variables and could only be presented by eight sub areas in

comparison to the 58 possible with the 1851 census.
The problems involved in making comparisons over
longer periods thus require the acceptance of a less
rigorous approach, but nevertheless, such comparisons hold more promise than the multiplication of
further studies of ecological change in the limited
period between 1851 and 1871.

American factorial ecologies of modern cities
have identified separate dimensions which correspond
quite closely to the social-area constructs of socio-
economic, family and ethnic status (Shevky and Bell,
1955; Johnston, 1976), but British studies have not
found such a good fit with theory.   The role of
housing conditions and tenure has emerged more
strongly in British studies, reflecting the tri-
partite division of the British housing market into
local authority, privately-rented and owner-occupied
sectors, and the association of each tenure with
particular parts of the city and particular types of
people (Murie, et al. 1976).   If it is true that
housing provision distinguishes post-war British
from American cities, we should expect a much closer
fit between the two in the years immediately preceding the First World War, when the social and
economic structure of the city had evolved sufficiently to warrant the description 'modern', but the
provision of housing remained the preserve of the
private landlord.   It is interesting that Pritchard
(1976) designates Leicester as an 'ecological city',
conforming to the models of Burgess and the classic
human ecologists, during the period 1870-1914, when
the structure of the housing market was dominated by
purpose-built, unfurnished privately-rented terraced
houses, with only an incipient sector of middle-
class owner-occupation to the south of the city
centre.

Langton and Laxton (1978) have urged that our
investigations should retreat to the eighteenth and
early-nineteenth centuries to discover the most
critical period of urban-industrial change, while
Colin Pooley's contribution to this volume notes the
dearth of studies of cities in the early-nineteenth
century.   My call on the other hand, is for us to
concentrate more effort on the late-Victorian and
Edwardian years.   This reflects a perspective
grounded in contemporary urban theory rather than
social and economic history, but that in fact is the
way in which most geographers have approached the
nineteenth century.   The least that can be said in
favour of the proposal is that it represents a progression from 'urban geographers in search of a
source' to 'urban geographers in pursuit of a theory',

which seems a marginally more respectable way of going about things!

The main difficulty of applying social area analysis to the late nineteenth century is the lack of the absolutely critical ingredient: data on family and household structure.   In the absence of the manuscript census a variety of records may be used to elucidate the patterns of social rank and ethnic status at the end of the century: ratebooks, directories, health records, church registers; but none of them is useful in matters of household size, kinship and family structure.   Inevitably, therefore, we have to rely more on types of evidence which, to date, have been used merely to illustrate or expand arguments based on quantitative analysis: novels, diaries, personal reminiscences, newspaper records, Royal Commission reports and social inquiries.

A closer integration of analyses of the built environment and the social environment is essential. Too much geographical research has concentrated too narrowly on purely social aspects and while we have developed a highly sophisticated quantitative methodology it often lends a spurious degree of precision to data which, however accurate, are only a partial representation of reality.   That a wider perspective is tenable is indicated by at least two studies: Pritchard's (1976) attempt to integrate housing, social structure and residential mobility in Leicester, and Daunton's (1977) even more wide-ranging survey of the economic and social structure of Cardiff.   Explanations based on the generalities of social area theory - technological change, industrialization, the rise (or fall) of class consciousness, the improvement of public transport, the expansion of the housing market, the changing roles of women and the family - must be replaced by detailed studies of how the opening of larger factories or more capital-intensive businesses, or the unity of the labouring classes, or the introduction of the tram, or the implementation of housing legislation, or the activities of philanthropists, or the abolition of child labour, or the introduction of compulsory education, or the employment of women outside the home, affected patterns of residential segregation and mobility in particular cities.

Community structure

The internally homogeneous social area on which ecological theory has focused is only one of many types of 'neighbourhood' recognized by urban sociologists and may have little if any functional significance for its inhabitants (Blowers, 1973; Bell and Newby, 1976). A continuum as interesting as that from Sjoberg to Burgess runs from the 'locality-based social system' (Stacey, 1969), typified by the 'urban village' (Connell, 1973), to the 'non-place community' (Webber, 1964), characterized by loose-knit social networks and long-distance patterns of social interaction. A map of perceived 'urban communities', based on actual patterns of behaviour, may be of more relevance to urban historians than a map of precisely, but anachronistically, defined 'social areas'. It is usually assumed that the modernization of society was associated with an increase in the distance over which social interaction took place, and a decline in the similarity of the contact patterns of neighbours, as improved transport services facilitated longer journeys, including longer journeys to work. Neighbours were less likely to be workmates and reduced hours of work and improved rates of pay gave workers the time and the money to travel further.

It is true, as Warnes (1970) and Vance (1967) have shown, that residence and workplace were still closely linked in the mid-nineteenth century and that many industrial towns comprised relatively self-contained cells, each focused on a particular mill or foundry. But this pattern should not be regarded as a continuation of pre-industrial cellular structure, since many cells developed only in the nineteenth century, including some that were planned as industrial communities, for example the colonies described by Marshall (1968), on the outskirts of Lancashire cotton towns.

In practice many of these workplace communities may not have functioned as 'urban villages' in the sense of being socially self-contained and close-knit, simply because they contained a whole range of status groups from millowner to mill operative, including supporting populations of clergy, school-teachers and tradesmen, between whom social, as opposed to functional, interaction may have been limited. It can be argued, therefore, that the development of larger segregated social areas in the second half of the century facilitated short-distance social interaction between social equals and the creation of 'urban villages' in working-class areas. Confirmation of this hypothesis is provided by patterns of marriage and residential mobility in the second half of the century: in Huddersfield, working-class districts in the inner city stood out more clearly as self-contained interaction systems with respect to marriage in 1880 than they had done in 1850, when the majority of low-status households had been less segregated from the middle classes and only the Irish had been clearly segregated, both spatially and socially (Dennis, 1975); and in both Huddersfield in the 1850s and Leicester in the 1870s groups of adjacent districts formed migration systems that accounted for the majority of residential moves made by their inhabitants (Pritchard, 1976; Dennis and Daniels, 1981) (Fig.10.1).

One problem in using these data to identify spatially-confined communities lies in the coarseness of the spatial framework which must be adopted if sufficient observations (marriages or moves) are to be associated with each area to make the analysis statistically meaningful. We expect most communities to be very small in extent: no more than a single court, or the houses served by a corner-shop or beerhouse. But statistical analysis can identify only larger and less closed communities formed by the association of several enumeration districts, maybe 5000 inhabitants strong. Nevertheless, 'urban village' hypotheses are supported by widespread evidence of the 'clannishness' of migrant groups, not only the Jews and the Irish (Connell, 1970; Lobban, 1971; Pooley, 1977; Lees, 1979; Lewis, 1979), but also relatively short-distance migrants - from Lancashire villages to Preston, from Westmorland to Huddersfield, from the counties of West Wales to Merthyr (Ashworth, 1954; Anderson, 1971; Carter and Wheatley, 1978a, 1978b; Dennis and Daniels, 1981) - and of the limited migration fields of the poor (Booth, 1892; Pooley, 1979).

Behavioural studies

An important theme in contemporary urban studies has been the reconciliation of individual and aggregate approaches to socio-spatial structure. Given that the first part of this chapter has lamented the in-adequacy of process-oriented explanations of residential patterns, it is appropriate that this section should focus on aspects of individual behaviour such as residential mobility and social interaction that may have provided mechanisms for social-area change. In this case individual behaviour is studied during a single period of time to throw light on the changes that occurred in ecological structure during that period. It is not

262

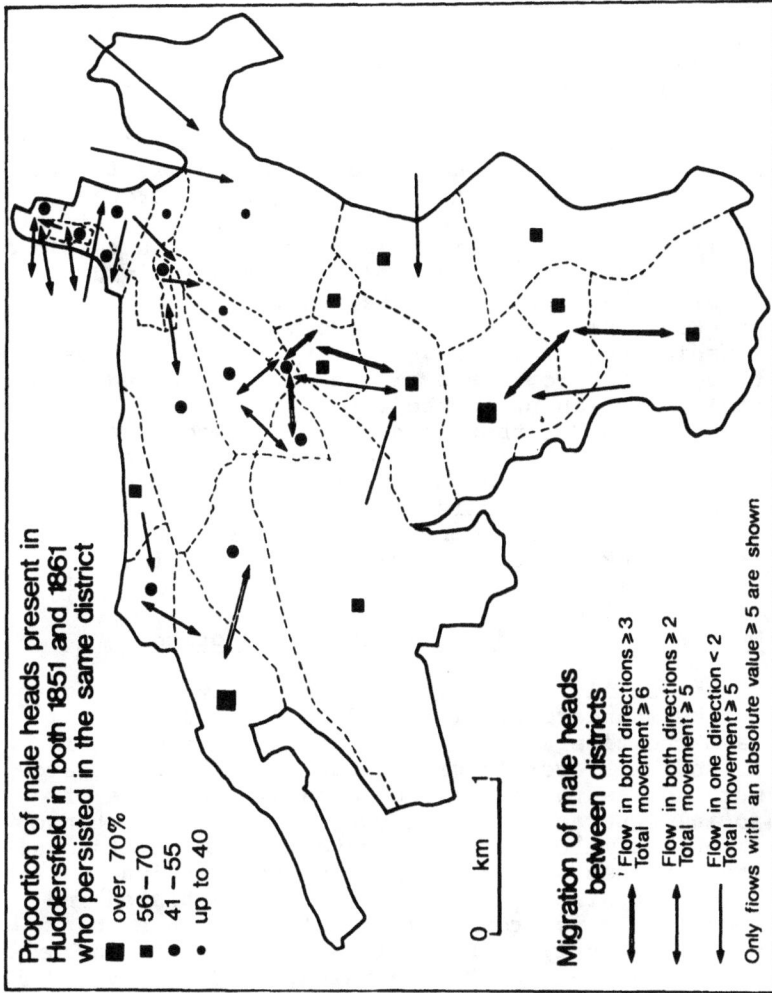

Proportion of male heads present in Huddersfield in both 1851 and 1861 who persisted in the same district

- ■ over 70%
- ■ 56 – 70
- ● 41 – 55
- • up to 40

Migration of male heads between districts

Flow in both directions ≥ 3
Total movement ≥ 6

Flow in both directions ≥ 2
Total movement ≥ 5

Flow in one direction < 2
Total movement ≥ 5

Only flows with an absolute value ≥ 5 are shown

0 ___ 1
km

Figure 10.1 Intercensal
mobility between enumeration
districts in south-west
Huddersfield, 1851-1861

263

hypothesized that the pattern of behaviour themselves changed through time, merely that their continued operation generated ecological change.

Another approach to behavioural studies is to consider how behaviour itself changes through time. Do the reasons for, or the patterns of, residential mobility change through time? Do individual fields of social interaction change in extent through time? To the extent that the context in which individual decisions are taken is itself changing as a result of earlier decisions, the ecological and individual levels of analysis are inseparable. But in practice, research on marriage patterns, migration, kinship and the journey-to work has often focused on these topics for their own sake, considering changes in interaction or in the extent of contact fields through time, rather than explaining aggregate change as a result of these individual decisions. Both approaches are important and, given the painstaking nature of this kind of research and its dependence on time-consuming methods of historical record linkage, both should be followed in individual research projects.

Marriage as a guide to social interaction

It is usually argued that as transport improves and as the standard of living rises individuals will have the time and the money to make and maintain friendships over longer distances. If we regard marriage as the outcome of a particular type of social interaction, then our hypothesis of change through time can be applied to marriage distances. The hypothesis of a decline in the frictional effect of distance over time has been tested and accepted in numerous historical situations, but most frequently in rural areas, e.g. southern Sweden, Dorset and central France (Rich, 1978; Perry, 1969; Ogden, 1973). An analysis of patterns of marriage in an urban area, Huddersfield, at three dates during the mid-nineteenth century, came to the expected conclusion that the poor married over shorter distances than the rich, but also arrived at the more unexpected conclusion that there was no significant difference between marriage patterns in 1850 and 1880 (Dennis, 1977a). This could be because the period of study was too short, but it could also be explained in terms of the changes in scale and degree of residential segregation that were discussed earlier. It was true that public transport by horse buses was better in Huddersfield in 1880 than it had been in 1850. But it was also

true that people were more likely to live in
segregated social areas in 1880, and therefore had
less need to travel far to find a sufficiently large
number of potential marriage partners, assuming that
most marriages were endogamous with respect to
status.   In rural areas the segregation theme was
irrelevant and so it is not surprising that the effect
of improved transportation was more obvious:
marriage distances did increase through time.   But
in both rural and urban areas another important
variable was population density.   The difference
between the marriage patterns of the rich and the
poor lay, not in the greater mobility of the rich,
but in the lower density at which they lived.   This
lower density necessitated searches over more
extensive areas to meet the same range of eligible
members of the opposite sex.

In this example, therefore, the effect of
distance on social interaction depended upon the
prevailing socio-spatial environment: the degree to
which interaction occurred within or between classes
or occupational groups, and the degree to which
those classes or groups were already living in
segregated areas.   Alternatively we could argue
that marriage patterns indicated the class structure
of the population (Foster, 1974), and that the
object behind segregation was to facilitate within-
class marriage (Beshers, 1962).   The complexity of
the relationship between residential patterns and
marriage distances is a particular example of the
argument developed by David Cannadine in chapter 9
on the interaction between shapes on the ground
(segregation) and shapes in society (in this case,
patterns of marriage).

Of course, the entire argument rests on the
suitability of marriage data as an indication of
social interaction, but we may doubt whether
marriage was the same type of contract for young as
for old, for rich as for poor, for immigrants as for
locally-born, or whether marriage was the same type
of contract in 1900 as in 1800.   For example, what
proportions of marriages were 'arranged' at
different periods or among different social groups?
How widespread was the notion of 'romantic love' as
the basis for marriage?   Although marriage
certificates yield information which appears to be
identical for all times and all peoples, such data
may actually hold very different meanings in
different circumstances.   We may acquire clues to
the role of marriage in Victorian society by considering literary evidence, contemporary tracts and
homilies on married life, diaries and - for the late

nineteenth century - the oral accounts of our
great grandparents. Similarly, some insight may be
derived from statistical evidence on age at marriage,
age-difference between partners, or place of
marriage (i.e. Anglican church, nonconformist chapel
or Register Office) associated with different
periods and groups. In Huddersfield the proportion
of marriages solemnized in nonconformist chapels increased from less than one in eight in 1851 to more
than a quarter by 1880, while the Register Office
increased its share of weddings from 2 per cent to
18 per cent in the same period. Average age at
marriage was 26.5 for grooms and 24.5 for brides in
Huddersfield in 1851, and 27 for grooms, 25.5 for
brides in 1880. In 1851 at least 18 per cent of
brides were older than their grooms, and at least 54
per cent of grooms were older than their brides. In
1880 these proportions were 17 per cent and 48 per
cent. None of this information on age implies any
great change in the characteristics of persons
getting married and this supports the validity of
comparing patterns through time.

Another complicating factor concerns the
relationship of marriage to economic conditons. We
are aware that the marriage rate fluctuated in
response to economic circumstances (Brown, 1978),
but we require further research on the degree to
which such fluctuations were concentrated among
particular social or age groups, and the effect of
any such bias on marriage distances. If the rich
were unaffected by economic problems, in a slump
the total number of marriages might be expected to
decline, but the number of 'rich marriages' to
remain constant and the proportion of 'rich
marriages' to increase. Given the tendency of the
rich to marry over longer distances than the poor,
the mean marriage distance would also increase,
although there would have been no real change in the
interaction behaviour of the population.

Finally, while trends in marriage patterns may
replicate those in other forms of informal social
interaction, it is not known whether average
marriage distances exceeded the distances over
which other contacts were maintained: the figures
have comparative but not absolute significance as
indicators of interaction.

Church membership

Another guide to day-to-day journey patterns of
Victorian townsfolk is the distance they travelled
to church. We can investigate changes in

geographical patterns of church membership through time, and again the major difficulties are those of interpretation rather than measurement. Not only is there the problem that 'membership' may be regarded differently by different denominations, but also we are dealing with a form of interaction regarded in different ways by different members of the same institution. Some members, who valued the church for its companionship as much as for its theology, may have been happy to attend the nearest church, whatever its denomination; others, for whom doctrine was more important, may have travelled distances to church that were quite unrepresentative of their other forms of interaction. Given the proportional decline in church attendance that occurred during the century, we might explain any increase in the journey-to-church distance not in terms of improved mobility, but as a reflection of the continued membership of only the most committed worshippers, who were prepared to travel long distances to the church of their choice.

Since church registers rarely include any information other than the names and addresses of members, it is difficult to control for age or status unless the mammoth task of linking church and census records is undertaken. But despite their shortcomings, church membership records (and the membership lists of other institutions, e.g. friendly societies, scientific or philosophical societies) can provide additional evidence of the community structure of Victorian cities (MacLaren, 1967; Rogers, 1972; McLeod, 1974, 1977; Dennis and Daniels, 1981). Links between the addresses of members and the location of their church may be used to identify territorially-defined communities and their changing configuration through time. Where occupational, age or birthplace information is included in listings (or can be obtained by record linkage with the census) membership data can also be used in the same way as marriage registers to identify social groupings in the population. The hypothesis that changing residential patterns represent attempts to adjust the spatial structure of the city to changing social structures can then be tested.

In practice, completed analyses of church membership rolls have been more modest in their scope. Analysis of nonconformist records in Huddersfield (membership rolls, lists of seat holders, Sunday school registers and baptism and burial registers) suggests that there was little overall change between 1850 and 1880 in the distances that people were prepared to travel to attend church. City-centre churches usually attracted more scattered congregations than suburban chapels, but in 1880 there were no places of worship for which records have survived where even half the members travelled more than one kilometre to attend.

The records of one denomination, the Independent/Congregationalist church, illustrate the variety of experience that is hidden by aggregate statistics. In 1851 Independent chapels were located near the centre of the town, in Ramsden Street, close to the traditional area of middle-class professional residence, and at Highfield, towards the inner end of a developing middle-class suburban sector. The records of Highfield chapel show that just over half its members lived within a kilometre and that 60 per cent lived closer to Highfield than to Ramsden Street. This situation continued until the 1860s when suburban congregational churches were founded in three areas of relatively high population density, but some distance from the two central churches. At first these new churches only drew support from their immediate neighbourhoods: in 1871 at least 80 per cent of their members lived less than one kilometre from their place of worship, but by no means all the Congregationalists living in those suburbs transferred their allegiance from the city-centre churches to their local church. The opening of suburban churches did not result in a contraction of the area served by the central churches, perhaps indicating the declining frictional effect of distance and the trend towards a 'non-place community'. By the end of the century suburban churches had also acquired a more dispersed membership, perhaps as members moved house but continued to worship at the church nearest their previous residence, or made recruits among their colleagues at work who lived in other parts of the city. Most Congregationalists belonged to the new lower-middle class (Crossick, 1977) and the emergence of non-conformist congregations as potential non-place communities in no way contradicts the continuing existence of 'urban villages' in working-class areas. Moreover, these data could equally well be interpreted as showing how slow 'locality social systems' were to disperse. What is clear is the need and the potential for further research.

Kinship

A third index of the distance over which social

268

interaction was regularly maintained is the
residential propinquity of related families, a
subject that attracted Anderson's attention in his
study of Preston (1971), and revived by Mills (1978)
in the rural context provided by the village of
Melbourn in Cambridgeshire.   Studies of this type
require elaborate record linkage of manuscript
censuses, maps and birth/baptism and marriage
registers.   Mills used the 1841 census (the least
satisfactory for this type of study since it omits
information on 'relationship to the head of household') but supplemented it with a family reconsti-
tution covering the previous half-century.   In a
more restricted analysis, using successive manuscript censuses for part of Huddersfield, a sample
of male household heads in 1861 was traced back to
1851 (Dennis, 1977b).   The sample included many
young married men who had been living in their
father's household ten years earlier.   The
addresses of 97 father-son pairs could be traced in
both censuses.   In 1861, 15 sons still lived in or
adjacent to their father's home, while 51 were now
living in a different enumeration district from
their father, but perhaps closer to their in-laws.
Only five had moved more than a mile away.
     Interesting as this information is, there are
substantial problems in its interpretation.
Firstly, the sample which could be traced must have
been biased towards stable and immobile families.
Furthermore, it is difficult to assess how far the
observed pattern differed from what might have been
expected on theoretical grounds.

Residential mobility

Perhaps the most fruitful field for the reconciliation of individual behaviour and ecological
structure is the study of residential mobility.
This has appeared as a subsidiary theme to social
mobility, usually in the form of 'turnover' or
'persistence' rates, in numerous American social
histories (Thernstrom, 1964; Knights, 1971; Katz,
1972), and more centrally in an increasing number
of British studies (Holmes, 1973; Pritchard, 1976;
Daunton, 1977; Dennis, 1977b; Pooley, 1979;
Dennis and Daniels, 1981).   At least three
approaches are possible:
(1) the interpretation of residential mobility in
     terms of choice-oriented theories applied to
     contemporary migration decisions, for example,
     the association of migration with age, status,

269

social mobility and the life cycle (Rossi, 1955;
Simmons, 1968; Brown and Moore, 1970);
(2) the analysis of mobility in terms of the operation of the housing market: the cost and ease of
moving, institutional constraints to mobility,
and the availability of vacancies (Robson, 1975;
IBG, 1976);
(3) the analysis of gross and net flows of migrants
between areas as the mechanism of ecological
change.
The only published study to compare mobility at
successive dates has been by Pritchard (1976), and
his research went little further than a comparison
of turnover rates after 1870. He explained the
long-term decline in the rate in terms of changes in
vacancies (their number decreased between 1870 and
1914), changes in tenure (less privately-rented
accommodation with the passage of time) and changes
in population structure (an increase in the number
of elderly heads of households). But Pritchard did
not elaborate on changes in the reasons for moving,
search processes or distances moved, all topics
worthy of further research.
The third approach, relating mobility to
changes in spatial structure, is the most relevant
to the present discussion. In fact, very little
evidence has yet been provided on the operation of
invasion and succession in Victorian cities. Dyos
(1961) noted that in Camberwell in the 1860s the
population of the East End of London moved frequently but clung to sets of adjacent streets as
they would to a country village; Lawton and Pooley
(1975a) recorded the frequent moves of David
Brindley, a dock railway porter in 1880s Liverpool,
who moved 12 times in 8 years, but never more than
4.2 km. and always within 2 km. of his workplace.
In Berry Brow, a suburb of Huddersfield, the
majority of male household heads in 1861 had been
living in the same area, if not the same house, ten
years earlier (Fig. 10.2; Table 10.1). Quantitative
analyses by Anderson (1971) in 1850s Preston,
Daunton (1977) in 1880s Cardiff and Pritchard (1976)
in 1870s Leicester indicate very high rates of
mobility, but much of it over short distances.
Pritchard estimated that one in every 4.5 households
moved each year in Leicester, but in many of the 26
areas into which he divided the city more than 40
per cent of intra-urban migrants moved to another
house in the same area. Anderson found that only
14 per cent of males in sample districts in Preston
had been living in the same house ten years earlier,
but another 26 per cent were traced to addresses

270

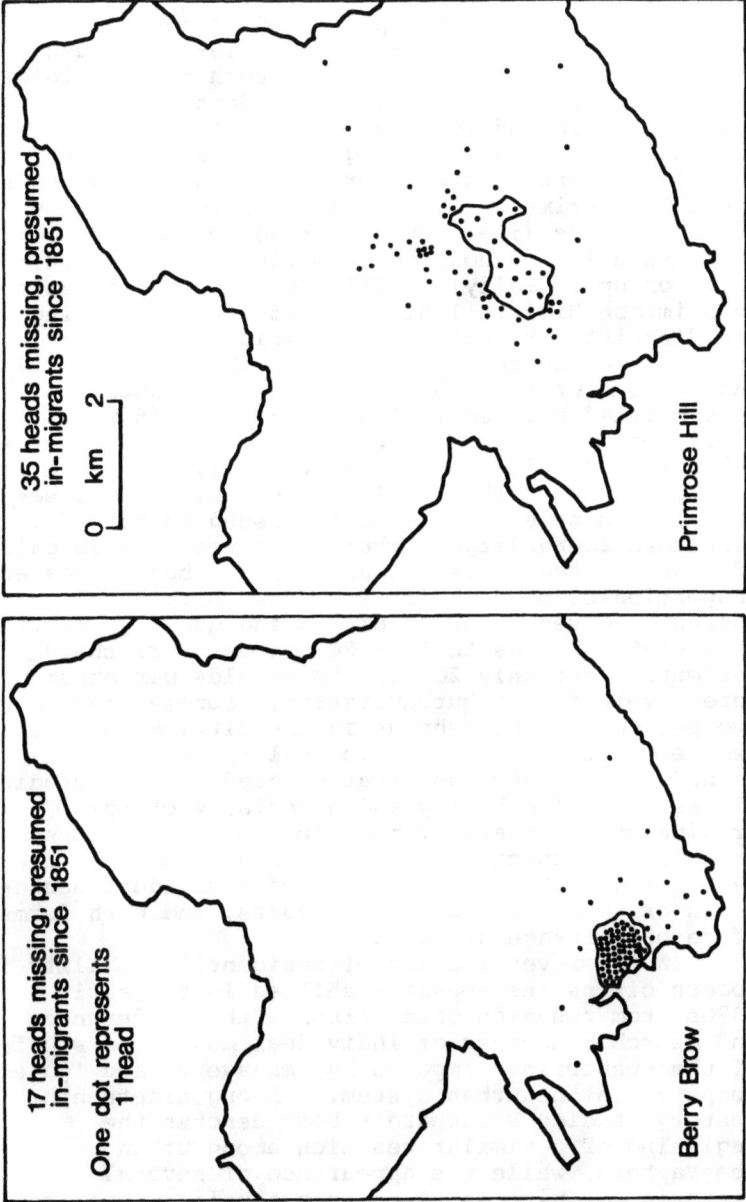

Figure 10.2 1851 residence of male household heads resident in Primrose Hill and Berry Brow in 1861

less than 200 yards away.   In Huddersfield
approximately 35 per cent of male household heads
remained in the same enumeration district from one
census to the next, while in Cardiff later in the
century only 13 per cent persisted at the same
address over a 10-year period, but 25-35 per cent
remained in the 'same area'.   In both Huddersfield
and Leicester it was possible to identify migration
systems linking adjacent districts, between which
migration was equally strong in both directions.

In Huddersfield a major area of new housing in
the 1850s, Primrose Hill, was occupied in 1861 by
new households (i.e. young married couples) as much
as by established households moving out geographically or up socially.   Of 103 male heads resident
in Primrose Hill in 1861, only 15 had been living in
the district ten years earlier while 53 could be
traced to an address elsewhere in the town (Fig.10.2;
Table 10.1). Of these 53 intra-urban migrants, 22
were married between 1851 and 1861, and 46 were
still less than fifty years old in 1861.   There is
only limited evidence in Huddersfield, therefore, of
a process of suburbanization by established households.   Considering only established households,
Pritchard found large numbers of moves between pairs
of houses rated at less than £5 p.a., but a greater
proportion of moves to or from more highly-valued
houses involved a change of housing quality, e.g.
from £5-£10 houses to £10-£20 houses.   Pritchard
estimated that only 200-300 households per annum
were involved in suburbanization, representing only
two per cent of households in the city, but a vital
two per cent as far as ecological change was concerned.   It is evident that we need further studies
of residential mobility and especially of the
origins of occupiers of new suburbs.   The process of
invasion and succession or filtering can be studied
by tracing either the fortunes of individual households or those of individual houses, and both forms
of record linkage are vital.

In the investigation of residential mobility in
modern cities the emphasis shifted in the early
1970s from research that stressed the preferences
and search behaviour of individual movers to studies
of the constraints imposed by 'managers' and 'gatekeepers' in the urban system.   Among nineteenth-
century studies Pritchard's book denotes the
beginning of a similar reaction among urban
geographers, while the appearance of several
histories of nineteenth-century housing provision
(Chapman, 1971; Tarn, 1973; Sutcliffe, 1974; Simpson
and Lloyd, 1977; Burnett, 1978) seem likely to

272

stimulate further interest in the geography of
housing in Victorian cities.

At present, remarkably little is known about
the activities of private landlords, their
selection of tenants, their methods of rent-fixing
and rent-collection, or the frequency with which
they evicted tenants.   In Cardiff 70 per cent of

Table 10.1: Intercensal Mobility in Huddersfield, 1851-61

Address in 1851	Resident in Berry Brow, 1861		Resident in Primrose Hill, 1861	
	All Heads	Heads Already Married in 1851	All Heads	Heads Already Married in 1851
Resident in same enumeration district	60	40	15	12
Elsewhere in Huddersfield	22	10	53	30
'Missing'	17	-	35	-

For the location of Berry Brow and Primrose Hill, see
Figure 10.2.
Source: Census Enumerators' Books, 1851 and 1861

houses were owned by investors with less than 10
houses, 4 per cent by companies and less than 10
per cent by owner-occupiers (Daunton, 1977).   But
these proportions varied widely between districts,
and owner-occupancy was more important in areas of
skilled working-class residence, where there was a
tradition of self-help, and in areas of higher-
rated housing.   We know very little about the
market for owner-occupied housing or about the
differences between the policies of large and small
landlords.   Nor, in larger cities, do we have much
direct evidence of the allocation policies in
pioneer council housing schemes or philanthropic
dwellings, apart from the frequently repeated
statement that the employment of the casual poor
was too irregular or too poorly-paid for them to
afford the rents asked for model dwellings, which
consequently became the preserve of the respectable,
regularly-employed working classes (Wohl, 1977).
The Victorians talked a great deal about 'levelling

up', but the effectiveness of this process has been doubted more often than it has been refuted by recent historians of working-class housing (Gauldie, 1974).

Much has been written about the displacement of the poor by central-area improvement, railway or road construction, or philanthropic housing schemes (Kellett, 1969; Wohl, 1977) but their movement has rarely been related to spatial changes in the structure of cities. Can we produce direct evidence of their crowding into adjacent districts and the subsequent deterioration of those districts? Can we trace the movement of individual households displaced by slum clearance schemes, and, equally important, can we trace the origins of the residents of the new housing that replaced the slums? Certainly, there is a case for applying a version of Rex's theory of housing classes to the late nine-teenth-century city (Rex and Moore, 1967; Rex, 1978). What was the relationship between the quality and tenure of housing that a family could obtain and the social or ethnic status of the family? And how did access to housing impinge on other aspects of co-operation or conflict among city dwellers such as voting patterns, society or church member-ship, intermarriage and the whole web of relation-ships that contributed to the city's community structure?

To answer any of these questions will require a mammoth task of record linkage involving censuses, ratebooks, directories, electoral rolls, pollbooks, membership lists, housing deeds, sale records and maps. Many of the problems of historic record linkage have been discussed by Wrigley (1973) and Morris (1976), but linkage through time in pursuit of residential or social mobility raises additional difficulties. Most linkages use occupation and/or address for identifi-cation, but in linkage through time changes in occupation and address are to be expected. Indeed, these are the changes in which we are most inter-ested. Intercensal linkage is still feasible, since a wide range of information on forenames, age, birthplace and the characteristics of other members of the same family can be used as identifiers, although the use of the last type of information biases linkage in favour of those with large families. Unmarried adults, especially if they have common names, may prove difficult to identify.

Census linkage is useful for distinguishing between long-term 'stayers' (those who do not appear to have moved at all between censuses, although

274

even these people may have moved out and back in
the intervening decade) and the rest, who may have
moved any number of times.   It is also useful in
studies of ecological change, where the net movement
of a household over a long period is more important
than frequent minor adjustments in its location and
circumstances.   But for other purposes the census
provides too infrequent a record of location, thus
requiring the use of successive directories,
electoral rolls or ratebooks, where the number of
identifiers is less and the chance of making an in-
correct link correspondingly greater.   As Morris
(1976) pointed out, sampling is undesirable since
it drastically reduces the chances of making links,
but some form of stratified sampling may be
acceptable, e.g. selecting particular surnames or
names beginning with certain letters of the
alphabet.

Some final thoughts

Much of the above may seem to have been unduly
negative.   The utility of some models of spatial
change has been doubted, the practicability of
testing other theoretically more interesting models
questioned, and the methods followed by many recent
quantitative analyses at both individual and
aggregate scales have been criticized.   More
positively, the following suggestions are made for
future research.
(1)   There is a need for a wider theoretical horizon
      for research on changes in residential
      differentiation than 'Sjoberg to Burgess'; to
      view the nineteenth century as a transitional
      period requires a clearer idea of the true
      nature of pre-industrial and twentieth-century
      cities;
(2)   Investigation is required of forms of spatial
      structure at a more detailed scale than
      'social areas', particularly the role of
      'villages within the city' and the perception
      of neighbourhood boundaries as revealed by
      what residents said and wrote, and how they
      behaved;
(3)   Greater emphasis should be given to the under-
      standing of historical processes rather than to
      the quantitative analysis of their expression
      which can only be as precise as the data are
      accurate, and as complete as the data are
      representative;
(4)   An integration of economic, morphological,

ecological and behavioural studies should be
attempted in the context of particular cities,
so that progress can be made beyond the
generalities by which spatial patterns have so
far been 'explained'. Ideally all the
evidence on processes of change should come
from the city in question, rather than by
assuming that it resembles other places where
transport provision, industrial organization or
the housing market have been investigated.
Perhaps it is time to recall Asa Briggs'
observation that: "The first effect of early
industrialization was to differentiate English
communities rather than to standardize them...
A study of English Victorian cities, in
particular, must necessarily be concerned with
individual cases" (Briggs, 1968, pp.33-4).

(5) Finally, there is a need to broaden the period
over which change is investigated. We need to
focus on the early nineteenth-century development of factory and mill employment, its
effect on the organization of the family and
the relationship between residence and workplace, and on the late nineteenth-century
improvement of mass transportation and increase
in local and central government intervention in
the provision of housing. Inevitably, such
research will involve a release from the ties
of the census enumerators' books and the use of
a much more diverse collection of sources than
has hitherto been common in geographical
studies of nineteenth-century towns.

Whatever forms of investigation we attempt, we
must be clearer than we have sometimes been in the
past as to why we are embarking on each new piece of
research, what theoretical framework we are working
within, and how our studies of individuals can have
wider significance for our understanding of the
communities to which they belong.

References

Aba-Lughod, J. (1969) 'Testing the theory of social area
analysis: the ecology of Cairo'. *American Sociological Review*,
34, 198-212

Anderson, M. (1971) *Family structure in nineteenth-
century Lancashire,* Cambridge University Press, Cambridge

Armstrong, W.A. (1966) 'Social structure from the early
census returns' in E.A. Wrigley (ed.), *An introduction to
English historical demography,* Weidenfeld and Nicolson,
London, pp.209-37

Armstrong, W.A. (1972) 'The use of information about occupation' in E.A. Wrigley (ed.), *Nineteenth-century society,* Cambridge University Press, Cambridge, pp.191-310

Ashworth, W. (1954) *The genesis of modern British town planning,* Routledge and Kegan Paul, London

Bell, C. and Newby, H. (1976) 'Community, communion, class and community action' in D.T. Herbert and R.J. Johnston (eds.), *Spatial perspectives on problems and policies,* Wiley, London, pp.189-207

Beshers, J.M. (1962) *Urban social structure,* Free Press, Glencoe, Illinois

Best, G. (1971) *Mid-Victorian Britain 1851-75,* Weidenfield and Nicolson, London

Blowers, A. (1973) 'The neighbourhood: exploration of a concept' in *The city as a social system,* Open University Press, Bletchley, pp.49-90

Booth, C. (ed.), (1892) *Life and labour of the people in London, Volume 3,* Macmillan, London

Briggs, A. (1968) *Victorian cities,* Penguin, Harmondsworth

Brown, G. (1978) 'Marriage data as indicators of urban prosperity', *Urban History Yearbook,* pp.68-73

Brown, L.A. and Moore, E.G. (1970) 'The intra-urban migration process: a perspective', *Geografiska Annaler,* 52B, 1-13

Burgess, E.W. (1925) 'The growth of the city' in R.E. Park (ed.), *The city,* Chicago University Press, Chicago, pp.47-62

Burnett, J. (1978) *A social history of housing 1815-1970,* David and Charles, Newton Abbot

Carter, H. and Wheatley, S. (1978a) 'Some aspects of the spatial structure of two Glamorgan towns in the nineteenth century', *Welsh History Review,* 9, 32-56

Carter, H. and Wheatley, S. (1978b) 'Merthyr Tydfil in 1851: a study of spatial structure', *Social and residential areas of Merthyr Tydfil: Working Paper* 2, U.C.W. Aberystwyth

Carter, H. and Wheatley, S. (1980) 'Residential segregation in nineteenth-century cities', *Area,* 12, 57-62

Chapman, S.D. (ed.), (1971) *The history of working-class housing,* David and Charles, Newton Abbot

Connell, J. (1970) 'The gilded ghetto: Jewish suburbanisation in Leeds', *Bloomsbury Geographer,* 3, 50-9

Connell, J. (1973) 'Social networks in urban society' in B.D. Clark and M.B. Gleave (eds.), *Social patterns in cities,* IBG, London, pp.41-52

Crossick, G. (ed.) (1977) *The lower middle class in Britain 1870-1914,* Croom Helm, London

Daunton, M.J. (1977) *Coal metropolis, Cardiff 1870-1914,* Leicester University Press, Leicester

Dennis, R.J. (1975) 'Community and interaction in a Victorian city', Unpublished Ph.D. thesis, University of Cambridge.

Dennis, R.J. (1976) 'Community structure in Victorian

cities in B.S. Osborne (ed.) *Proceedings of British-Canadian Symposium on Historical Geography,* Queen's University, Kingston Ontario, pp. 105-38

Dennis, R.J. (1977a) 'Distance and social interaction in a Victorian city', *Journal of Historical Geography,* 3, 237-50

Dennis, R.J. (1977b) 'Intercensal mobility in a Victorian city', *Transactions of the Institute of British Geographers,* N.S. 2, 349-63

Dennis, R.J. (1979) 'Introduction' to 'The Victorian City', *Transactions of the Institute of British Geographers,* N.S. 4, 125-128

Dennis, R.J. (1980) 'Why study segregation? More thoughts on Victorian cities', *Area* 12, 313-7

Dennis, R.J. and Daniels, S.J. (1981)''Community' and the social geography of Victorian cities', *Urban History Yearbook,* pp.7-23

Dyos, H.J. (1961) *Victorian suburb,* Leicester University Press, Leicester

Foster, J. (1974) *Class struggle and the industrial revolution,* Weidenfeld and Nicolson, London

Gauldie, E. (1974) *Cruel habitations,* Allen and Unwin, London

Harris, C.D. and Ullman, E.L. (1945) 'The nature of cities', *Annals of the Academy of Political and Social Sciences,* 242, 7-17

Harrison, J.F.C. (1971) *The early Victorians 1832-51,* Weidenfeld and Nicolson, London

Holmes, R.S. (1973) 'Ownership and migration from a study of rate books', *Area,* 5, 242-51

Holmes, R.S. and Armstrong, W.A.(1978)'Social stratification', *Area,* 10, 126-8

Institute of British Geographers (1976) 'Houses and people in the city', *Transactions of the Institute of British Geographers,* N.S. 1

Johnston, R.J. (1972) 'Towards a general model of intra-urban residential patterns' *Progress in Geography,* 4, 82-124

Johnston, R.J. (1976) 'Residential area characteristics: research methods for identifying urban sub-areas', in D.T. Herbert and R.J. Johnston (eds.), *Spatial processes and form,* Wiley, London, pp.193-235

Katz, M. (1972) 'The people of a Canadian city: 1851-2'. *Canadian History Review,* 53, 402-26

Kellett, J. (1969) *The impact of railways on Victorian cities,* Routledge and Kegan Paul, London

Knights, P.R. (1971) *The plain people of Boston, 1830-1860,* Oxford University Press, New York

Langton, J. (1975) 'Residential patterns in pre-industrial cities'. *Transactions of the Institute of British Geographers,* 65, 1-27

Langton, J. and Laxton, P. (1978) 'Parish registers and urban structure: the example of late eighteenth-century Liverpool', *Urban History Yearbook,* pp.74-84

278

Lawton, R. and Pooley, C.G. (1975a) 'Individual appraisals of nineteenth-century Liverpool', *Social geography of nineteenth-century Merseyside project: Working Paper 3*, University of Liverpool

Lawton, R. and Pooley, C.G. (1975b) 'The urban dimensions of nineteenth-century Liverpool', *Social geography of nineteenth-century Merseyside project: Working Paper 4*, University of Liverpool

Lawton, R. and Pooley, C.G. (1976) *The social geography of Merseyside in the nineteenth century*, Final report to S.S.R.C.

Lees, L.H. (1979) *Exiles of Erin: Irish migrants in Victorian London*, Manchester University Press, Manchester

Lewis, C.R. (1979) 'A stage in the development of the industrial town: a case study of Cardiff, 1845-75', *Transactions of the Institute of British Geographers*, N.S. 4, 129-152

Lobban, R.D. (1971) 'The Irish community in Greenock in the nineteenth century, *Irish Geography*, 6, 270-81

McElrath, D.C. (1968) 'Societal scale and social differentiation: Accra, Ghana' in S. Green (ed.), *The new urbanization*, St. Martin's Press, New York, pp.33-52

MacLaren, A.A. (1967) 'Presbyterianism and the working class in a mid-nineteenth century city', *Scottish Historical Review*, 46, 115-39

McLeod, H. (1974) *Class and religion in the late Victorian city*, Croom Helm, London

McLeod, H. (1977) 'White-collar values and the role of religion' in G. Crossick (ed.) *The lower middle class in Britain 1870-1914*, Croom Helm, London, pp.61-88

Marshall, J.D. (1968) 'Colonisation as a factor in the planting of towns in north-west England' in H.J. Dyos (ed.), *The study of urban history*, Arnold, London, pp.215-30

Mills, D.R. (1978) 'The residential propinquity of kin in a Cambridgeshire village, 1841', *Journal of Historical Geography*, 4, 265-76

Morris, R.J. (1976) 'In search of the urban middle class', *Urban History Yearbook*, pp.15-20

Murie, A. et al. (1976) *Housing policy and the housing system*, Allen and Unwin, London

Ogden, P.E. (1973) 'Marriage patterns and population mobility', *University of Oxford School of Geography Research Paper 7*

Peach, C. (ed.) (1975) *Urban social segregation*, Longman, London

Perry, P.J. (1969) 'Working-class isolation and mobility in rural Dorset, 1837-1936'. *Transactions of the Institute of British Geographers*, 46, 121-41

Poole, M.A. and Boal, F.W. (1973) 'Religious residential segregation in Belfast in mid-1969: a multi-level analysis' in B.D. Clark and M.B. Gleave (eds.) *Social patterns in cities*, IBG, London, pp.1-40

Pooley, C.G. (1977) 'The residential segregation of migrant communities in mid-Victorian Liverpool', *Transactions of the Institute of British Geographers*, N.S. 2, 364-82

Pooley, C.G. (1979) 'Residential mobility in the Victorian city', *Transactions of the Institute of British Geographers*, N.S. 4, 258-277

Pritchard, R.M. (1976) *Housing and the spatial structure of the city*, Cambridge University Press, Cambridge

Rex, J.A. (1968) 'The sociology of a zone in transition' in R.E. Pahl (ed.) *Readings in urban sociology*, Pergamon, Oxford, pp.211-31

Rex, J.A. and Moore, R. (1967) *Race, community and conflict*, Oxford University Press, Oxford

Rice, J.G. (1978) 'Indicators of social change in rural Sweden in the late nineteenth century', *Journal of Historical Geography*, 4, 23-34

Robson, B.T.(1971) *Urban analysis: a study of city structure*, Cambridge University Press, Cambridge

Robson, B.T. (1973) 'A view on the urban scene' in M. Chisholm and B. Rodgers (eds.), *Studies in human geography*, Heinemann, London, pp.203-41

Robson, B.T. (1975) *Urban social areas*, Oxford University Press, Oxford

Rogers, A. (1972) *This was their world: approaches to local history*, BBC, London

Rossi, P.H. (1955) *Why families move*, Free Press, Glencoe, Illinois

Royle, S.A. (1977) 'Social stratification from the early census returns: a new approach', *Area*, 9, 215-9

Schnore, L.F. (1965) 'On the spatial structure of cities in the two Americas' in P.M. Hauser and L.F. Schnore (eds.), *The study of urbanization*, Wiley, New York, pp.347-98

Shaw, M. (1977) 'The ecology of social change: Wolverhampton 1851-71', *Transactions of the Institute of British Geographers*, N.S. 2, 332-48

Shevky, E. and Bell, W. (1955) *Social area analysis*, Stanford University Press, Stanford

Simmons, J.W. (1968) 'Changing residence in the city: a review of intra-urban mobility', *Geographical Review*, 58, 622-51

Simpson, M.A. and Lloyd, T.H. (1977) *Middle-class housing in Britain*, David and Charles, Newton Abbot

Sjoberg, G. (1960) *The pre-industrial city*, Free Press, Glencoe, Illinois

Stacey, M. (1969) 'The myth of community studies', *British Journal of Sociology*, 20, 134-45

Sutcliffe, A. (ed.) (1974) *Multi-storey living: the British working class experience*, Croom Helm, London

Tarn, J.N. (1973) *Five per-cent philanthropy*, Cambridge University Press, Cambridge

Thermstrom, S. (1964) *Poverty and progress: social mobility in a nineteenth century city*, Harvard University

Press, Cambridge Mass

Timms, D.W.G. (1971) *The urban mosaic,* Cambridge University Press, Cambridge

Vance, J.E. (1966) 'Housing the worker: the employment linkage as a force in urban structure', *Economic Geography,* 42, 294-325

Vance, J.E. (1967) 'Housing the worker: determinative and contingent ties in nineteenth-century Birmingham', *Economic Geography,* 43, 95-127

Vance, J.E. (1971) 'Land assignement in the pre-capitalist, capitalist and post-capitalist city', *Economic Geography,* 47, 101-20

Vance, J.E. (1977) *This scene of man,* Harper and Row, New York

Ward, D. (1975) 'Victorian cities: how modern?', *Journal of Historical Geography,* 1, 135-51

Ward, D. (1976) 'The Victorian slum: an enduring myth?' *Annals of the Association of American Geographers,* 66, 323-36

Ward, D. (1978) 'The early Victorian city in England and America: on the parallel development of an urban image' in J.R. Gibson (ed.), *European settlement and development in North America,* Toronto University Press, Toronto, pp.170-89

Ward, D. (1980) 'Environs and neighbours in the "Two Nations": residential differentiation in mid-nineteenth-century Leeds'. *Journal of Historical Geography,* 6, 133-62

Warnes, A.M. (1970) 'Early separation of homes from workplaces and the urban structure of Chorley', *Transactions of the Lancashire and Cheshire Historic Society,* 122, 105-35

Warnes, A.M. (1973) 'Residential patterns in an emerging industrial town' in B.D. Clark and M.B. Gleave (eds.), *Social patterns in cities,* IBG, London, pp.169-89

Webber, M.M. (1964) 'The urban place and the nonplace urban realm' in M.M. Webber (ed.), *Explorations into urban structure,* University of Pennsylvania Press, Philadelphia, pp.79-153

Wohl, A.S. (1977) *The eternal slum: housing and social policy in Victorian London,* Arnold, London

Wrigley, E.A. (ed.) (1973) *Identifying people in the past,* Arnold, London

Chapter 11

INDICATORS OF POPULATION CHANGE AND STABILITY IN NINETEENTH-CENTURY CITIES: SOME SCEPTICAL COMMENTS

Michael Anderson

Introduction

There are clear variations in the extent to which the different social science disciplines consciously make use of a methodology in which theoretically inspired concepts are operationalized by explicitly specified indicators: nevertheless, explicitly or implicitly, such a methodological procedure underlies all social science approaches to empirical data.   In what follows it is assumed as uncontroversial that we are not interested in everything about the present or the past, but only in that information which throws light on some empirical controversy, or on some pressing social problem, or which forms an important element in some larger, theoretically-inspired, analysis.

This chapter is concerned with one kind of data - information on patterns of population turnover and stability - which is frequently used (though often implicitly) to 'indicate' a number of theoretically important concepts in the scientific study of social change.   It takes as its main theme an assessment of the validity and reliability of these indicators for the purposes for which they are used.   The chapter also attempts to evaluate the theoretical significance of the kinds of conclusions which can be drawn from these data.

In the recent past, the kind of study reviewed here has most frequently been pursued by two groups of social scientists - social historians (e.g. Knights, 1971; Thernstrom and Knights, 1971; Chucadoff, 1972; Thernstrom, 1964, 1973; Scott, 1969; Katz, 1975; Daunton, 1977) and historical geographers (e.g. Holmes, 1973; Pritchard, 1976; Dennis, 1977).   Typically the material used has been collected as part of a wider study and has been derived from sources consisting of two or more nominal listings of the population - most usually censuses, though directories, lists of electors and other nominal listings have been used. The entries in these sources have been compared over a period of time (varying from a few days to ten years) and note taken of the numbers of individuals who can be found in both listings (who can be 'linked') and of those who have entered or departed from the listings between the two dates. In fact, the linking procedures have varied significantly and this has led to major problems of reliable comparison (Katz, 1975, p.120). Information is also usually collected on the numbers who have been traced between the sources but who have, nevertheless, been recorded as having a different address in the two listings; usually some attempt is also made to trace the direction of movements.

Problems connected with the reliability, validity and theoretical significance of this kind of information can be cast into a number of specific questions. For example, an immediate question which arises in this. Does a knowledge of a rate of population turnover, population inflow, population outflow or of the distances that resident individuals moved provide a useful indication of anything about the social structure and social relationships of a town or of an area within it? If 'yes' what degree of difference between areas, and in which rates, is important? Over what kind of time-span should mobility be studied for optimal results? If distance is relevant, can we say anything in general about how far people must move for that movement to be considered socially significant? Also, at an individual level of analysis, is it relevant or useful knowledge about a person (i.e. does it tell us something about that individual's social relationships or attitudes or beliefs), to be able to say that he or she has or has not moved any particular distance or in any particular direction in any particular time period? Again, if the answer is yes, which of these pieces of information is useful, and how much and what kind of movement is of interest to us? Finally, on an ecological level of analysis, a further set of questions can be asked about the theoretical significance of data on individual movements for an understanding of the changing social structure of urban areas. These ecological questions are explored in the last section of the chapter.

Questions of this kind do not raise purely

284

abstract and hypothetical problems. Such questions
are important because we can only evaluate different
techniques by reference to them.   Only when such
issues are borne in mind can we weigh up the significance of omissions, the seriousness of the
problem of invalid linkages, and of 'lost'
individuals.   Similarly, only in this way can we
adequately answer the question of whether sources
which only allow the tracing of individuals over a
ten year period (censuses) are useful for practical
research purposes.   Of course, the answers to
these questions will vary somewhat according to
which problem is being investigated,but it is clear
that we have often been paying too little attention
to the question of why we measure what we are
measuring.   Here, as in many other areas of social
history, we have perhaps become just a little
mesmerized by the fact that we can measure anything
at all about the past with any degree of quantitative rigour, and have paid too little attention to
the meaning of what we have 'discovered' (c.f.
Daunton, 1977, p.141).

Population stability at the individual level

If, then, we turn attention to some of the
inferences which have been made from these sources
about nineteenth-century cities, we can observe a
range of sometimes contradictory conclusions as
well as a surprisingly large number of accounts
where information on population turnover is presented but treated as intrinsically interesting in
itself.   This is the case to a great degree in
Knights (1971), Thernstrom and Knights (1973) and
Chucadoff (1972).   In as far as the criticism is
also relevant to Dennis (1977), it also raises
similar questions to those raised here about the
proper focuses of migration theory.   At the
individual level, four particular lines of analysis
can be distinguished, all of which can be related to
two more general and dubious assumptions.   Firstly,
it is implied that population turnover involved
disruption of social relationships and that mobility
can be taken as some kind of indicator of the
strength of a sense of community and of the likely
availability of supportive family or other primary
relationships (Daunton, 1977, pp.131-42 who is very
influenced by the work of the Institute of Community
Studies and particularly Young and Willmott, 1962;
and Willmott, 1963; also Thernstrom and Knights,
1971).   Secondly, the level of mobility is seen to

285

be correlated with problems of social control both over public behaviour and over personal morality. A view which stressed problems of control was certainly held by some contemporaries (Pritchard, 1976, pp.63-4) and in general was influential in the older sociological literature and particularly that of the Chicago School. A contrary view, arguing for ease of control by elites over transient populations is found in Thernstrom (1973, pp.231-2). Thirdly, mobility is assumed to be related to the extent to which communities can meet the political and social needs of their members by the availability or otherwise of adequate numbers of individuals able to provide continued and stable incumbency of certain key social roles (Katz 1975, p.134). Fourthly, a direct correlation - though operating in either direction - has been posited between social stability and political or class consciousness (Scott, 1969; Griffen, 1969; Thernstrom and Knights, 1971; Chucadoff, 1972, summarized in Katz, 1975, pp.111-4).

Underlying these four propositions there seem to be two assumptions. The first is that mobility is directly correlated with disruption; the second (though perhaps less conscious) is that the resultant (assumed) problems which occur in cities with high levels of turnover in the nineteenth century are either new or are different from those of earlier urban communities because rates of movement in nineteenth-century cities were significantly higher. Both these assumptions must be questioned.

We can begin to think about some of these problems if we consider two ideal-type communities at opposite ends of a continuum. In one, no-one ever moves in or moves out; everyone dies in the community of their birth, having lived in the same house for their whole lives. In the other, population turnover is total and continuous; no-one (at all) stays for more than a few days before moving and both in-migrants and out-migrants travel long distances when they come to the community and when they leave it.

Clearly, in these two extreme cases, we should expect to find some differences in patterns of social relationships between the two communities. The first would, in fact, be ethnographically unique in that it would have either many very large households or universal incest! Either way the community would almost certainly change eventually because random age-sex imbalances in the population would gradually lead to the extinction of some residential groups while others would become too

286

large for their locally available resources. Thus, some degree of population mobility within the community is necessary to counteract random demographic effects and some, though a lesser, degree of movement into and out of the community is also necessary for the same reason. The larger the community, however, the smaller will be the need; community size, indeed, by affecting the degree to which random effects are suppressed, is highly relevant to much of what follows.

In differentiated societies a problem of lack of flexibility in social structure and production would also arise in our perfectly stable case. Thus, we can also conclude that some degree of population movement is essential in order to maintain proper adjustment in the labour supply available to the community (for example, a village might have a regular demand for one carpenter but would require more than one if major building works were to be undertaken). However, given nineteenth-century mortality and fertility rates, it seems likely that in all but the smaller rural communities fairly small annual movements would be enough to solve the problems just discussed. In most Victorian towns the random fluctuations would be largely ironed out and few problems would arise as long as the economic base did not change. (When it did, of course, some new labour would be required and it is a relevant question for the historian of labour supply to ask where the different categories of labour came from? This still largely unexplored problem is a very different one from that of the study of population stability *per se*.)

Another feature which we should expect to find in our perfectly stable population (or even in the minimally mobile modification just discussed) is that everyone would know a very large number of other people in the community. Indeed, if the community was of small or medium size everyone might actually know everyone else, and in the jargon of the network theorists we should then have a perfectly dense or close-knit network. While this is true, however, it is not so clear that anythink very interesting follows from it. Close-knit networks provide extremely effective media for the exercise of social control but how effective they are depends on the extent to which individuals can be independent of others if necessary (Anderson, 1971, ch.2). This is partly a function of the modes of production and political organization of the community (and thus is highly variable),

287

and partly a function of the extent to which
individuals are subjected to mediating constraints
imposed by external forces such as the State or the
Local Landlord.   It may also be noted that control
can operate almost equally effectively in
communities which are some way from totally dense
and that quite effective networks can be built up
in a few months or years by individuals with
suitable needs, motivations and social skills (see,
for example, Bell, 1968, ch.6).   The extent of
social control also depends on the visibility of
deviant behaviour and this, in turn, depends on the
possibilities for segmenting social relationships
into different social and spatial areas, on the uses
made of public space, and on the extent of norms
regarding the privacy of particular aspects of
behaviour.   All of these features vary between
communities,both at a point of time and over time
in ways unrelated to population turnover (Laslett,
1963; Shorter, 1976; Anderson, 1979).   In sum,
while the possibility of social control is
certainly maximized in extremely stable communities,
there is no necessary correlation even at an ordinal
level between control and population turnover.
Under certain circumstances, high levels of social
control and high levels of functional interdependence between individuals can both develop in very
short periods of time in communities with quite high
rates of turnover and where relationships are well
below maximum density.

If we turn attention now to the polar
opposite ideal type, the totally mobile community,
some clear problems arise both in social control
and in the filling of all but the most ephemeral
social roles.   For example, in this type of
community capital investment by the population
would be of negligible proportions and specialized
social leadership would be lacking.   If leadership
is lacking and all contacts ephemeral it would be
difficult for class organization to develop.
However, as soon as a small departure is allowed
from the extremity of the continuum and even a few
individuals are assumed to be resident for a
relatively long period, the implications for the
model can change completely if we also assume that
these individuals are economically and socially
active.   Indeed, this variable level of involvement on the part of permanently resident individuals reveals itself as being a crucial intervening
variable in this part of the analysis.   Admittedly,
if some individuals are stable in a mainly transient
population, significant inequalities in power are

likely to emerge. These inequalities result from
differential abilities to manipulate locally
available resources, but it is by no means inevitable that social control will break down and
that necessary community roles will go unfilled.
For example, Thernstrom (1964, pp.168-80, 1973,
p.232) has pointed out the importance of enduring
local institutions and voluntary associations in
maintaining social stability in highly transient
areas. Much obviously depends on who the stable
individuals are and many other intervening
variables are likely to operate between population
turnover and patterns of interaction at the
community level. An example of social control in
transient communities can readily be seen in hotels
where populations are generally even more transient
than in the polar ideal type, but where social
control very seldom totally breaks down. Indeed,
most of the problems of human resource availability
and social control in any large community are
likely to be more dependent on who is present in
the community in the first place than on who is
transient, and much more on who is transient than
on the amount of transiency as such (c.f. Katz,
1975, p.117).

So far, it has been implicitly assumed in this
analysis of transiency that the population is an
atomized one lacking in social connections between
members. This picture may correspond to the image
of central city areas held by the great urban
theorists of the 1930s (and even of the Victorians
themselves), but we now know that in nineteenth-
century cities there was considerable social order
in the slum and that this in part arose from the
pre-existing social relations of members. People
do not, in general, migrate as isolated individuals
but move along well charted migration paths and
these embed them into pre-existing urban networks
on arrival in the city. It is, indeed, the
presence of enduring patterns of social relationships of this kind, involving individuals of
relatively homogeneous ethnic and social backgrounds, that provide one of the major sources of
neighbourhood and ecological stability (Anderson,
1971; Katz, 1975, pp.107-8; Hareven 1975; Hareven
and Langenback, 1979). In nineteenth-century
cities such networks seem to have performed considerable social functions for their members and to
have exercised significant control over their
activities, as well as influencing residential and
locational patterns in the community at large.

We can conclude that both social control and

residential location are not simply a direct function of population turnover but are mediated in an *a priori* unpredictable manner by pre-existing social relationships. The effects of population turnover cannot be assumed but must be subjected to empirical verification; population turnover itself is an unreliable indicator of these effects. It is perhaps only certain extremely mobile individuals, who never in their lives settle for more than a short time in any one place, who necessarily have problems of social integration into new communities. Current research techniques, however, do not isolate this group, let alone begin to investigate the consequences of their transiency.

This leads to the second question posed earlier - the effects of migration on individuals. Today we tend to have a stereotyped image of house moving as an essentially disruptive activity. There are problems of disposing of old property and of acquiring new. There is the cost and time involved in organizing the transport of belongings, the furnishing of the new residence and the organization of the connection and disconnection of necessary services; and there may also be considerable social and emotional disruption. In the nineteenth century many of these problems would have been much less significant for the mass of the population. Most moves were short distance (often very short), people had relatively few belongings to move and few services to organize, property turnover was high and, since most property was rented, little commitment was involved if the new residence turned out to have unforseen drawbacks (Davies, 1963; Pritchard, 1976; p.183-4). Moreover, we should not assume that the personal costs of movement were seen as particularly high since, both in urban and rural areas, nineteenth-century British populations were highly mobile - in a sense transiency was part of the expected life-experience of most people, perhaps even part of the culture since high rates of mobility were present in all parts of the country, rural and otherwise.[1] What is very much lacking from our knowledge of population mobility in the past is enough information on the attitudes to movement and to instability held by those most intimately affected. Certainly there is some evidence that at least short-distance moves were treated fairly casually, rather like the fact of commuting to work in the modern city (Katz, 1975, pp.94-111; Davies, 1963; Pritchard, 1976, pp.183-4). If this is so, then in analysing and describing short-distance moves in great detail and with no little effort, we may be grossly over-estimating their significance. Certainly we should be on our guard against automatically imputing our own meanings onto nineteenth-century patterns of behaviour when their significance for contemporaries may have been quite different. The context and the functions of mobility as seen by individuals are crucial here, a point clearly emerging in the discussion of the relationship between population stability and class consciousness (Katz, 1975, pp.111-4).

A further implication of the points raised in the last paragraph is that we should be hesitant in making comparisons of population turnover covering long periods, since with the passage of time the functions, effects and symbolic significance of turn-over may have changed for those involved. Although we know that rates of mobility among the urban working class have fallen very considerably over the past 150 years (Pritchard, 1976; Thernstrom, 1973; Knights, 1971) we cannot draw any very useful inferences from this 'knowledge' until we know a great deal more about motives, attitudes and effects. If we simply compare rates we may well not be comparing like with like. In the same vein, the effects and significance of mobility are likely to be very different for different occupational and different ethnic groups with their very different cultural traditions of migration (McLaughlin, 1971; Lees, 1969). If we simply compare rates of mobility of large social groups or, worse, aggregate them together into some overall 'average' analysis, we may totally miss the significance of what we are actually observing, both at a point in time and over time.

We also need to treat very carefully assumptions that moves involve disruption of pre-existing social relationships. Given the size of nineteenth-century British cities outside London regular contact could still have been maintained with someone who had moved anywhere within the city, as long as motivation existed. Moreover, since walks of twenty miles a day to see relatives were not unknown, movement to the next town may often have been no more significant than movement to the other side of the same town, at least in areas of fairly dense urban settlement.

These considerations raise a general issue of some significance. In analysing the results of record linkage exercises the population has generally been categorized into three basic groups: (i) disappearers. (ii) movers, and (iii) stayers.

Within these three broad groups, populations have been further categorized by distance moved. While the distances involved may be an important variable in understanding the consequences of mobility, it is not clear what distances were relevant in terms of their effects on those involved, nor whether these socially-relevant distances remained constant over time. Indeed, the likelihood is that they were not. If we reconceptualize the problem in terms of socially-relevant distance, a number of points become clear. For example, if we always treat a single town as our unit of analysis, this may lead us to miss some of the most socially-significant moves. For example, one piece of Swedish research has clearly shown the importance of treating the town and its rural environment as one unit of analysis for this purpose, by showing the subtle changes in patterns of urban and suburban residence that occurred during the industrialization period (Ohngren, 1974; c.f. Knights, 1971, p.113). By treating someone who moves outside our area of study as inherently different from someone who moves within it, we may be making a quite false distinction in terms of the social relationships and the social attitudes of the people involved and may thus be reaching both invalid and theoretically sterile conclusions.

Similarly, while we habitually distinguish intra-urban movers from those who do not move at all, this too may be an irrelevant categorization. Indeed, the most important variable in any analysis of the disruptive effects of migration on social relationships is not distance but something very much more difficult to measure, the relative importance of local (neighbourhood) relationships compared with other relationships such as those of kinship and friendship which are not primarily based upon the immediate locality. The strength and the functions of these relationships are poorly correlated with distance. In the contemporary world, for example, the life chances of many middle class couples may be much more dependent upon their relationships with parents living hundreds of miles away than they are on contacts within the neighbourhood (Bell, 1968; Hubert, 1965). Also relevant when considering the effects of short-distance moves on social relationships is the extent to which these moves take people into different networks, so that their normal transport routes, shopping areas and leisure activity centres change and their access to their previously most functional and important social contacts changes with

292

them. If we are to be able to assess the effect of movement (and therefore know what kinds of movement it is valid and theoretically relevant to measure) we need to know much more than we have usually gathered in the past about neighbourhood activity spaces. We have to do this before we can interpret our analyses of population movement and, indeed, even before we can set up meaningful analytical categories, spatial or otherwise.

Does the analysis so far presented in this chapter therefore suggest that the study of short-term individual-level patterns of mobility is irrelevant to the concerns of the social historian? Probably not, but it does suggest that some of the more easily calculated rates are not of great interest to him. The most interesting questions demand much more complex, more multifaceted and more difficult research enterprises.

One area with which we should be more concerned is the analysis of the meanings of movement and, particular, of its functions and effects as they appeared to those involved. To do this would require that for long periods we should forget our computers and turn much more to the sources and methods of the 'traditional' social historian.

A second area which needs more attention is life-time patterns of mobility. It seems much more interesting to know who (in terms of social roles) moves, how many times they move and where they move, rather than to know how much movement there is within or out of an area in any particular period of time.

Thirdly, if we are to study movement within and out of areas we need both to ensure that our geographical spaces are also relevant social spaces and that our time spans are not too long to vitiate our conclusions. Within ten years in many areas today most people come to be treated as locals, and in the nineteenth century, when ten years could be half of an adult's prime lifetime and when population turnover was so high, this seems even more likely to have been the case. Moreover, there is no necessary correlation between ten-year movement and movement rates over shorter periods of time.

Fourthly, given the poor validity and reliability of most measures of turnover as indicators of important social situations, we must beware of being mesmerized by relatively small differences in rates or in distances moved. Only easily noticeable differences are likely to be of unambiguous significance.

Secular stability and change

In this final section attention is turned to only one question - does the information which we can gather on population movements for short periods contribute to our understanding of secular processes of change in urban areas? In theory, of course, the answer is 'yes' since changes in the social composition of the population of different areas of cities must result in great part from net differences between the social characteristics of the in-migrating and the out-migrating populations. The previous sentence read 'in great part', because life cycle effects are also significant in making possible an ageing of an area's population even in the absence of movement (with consequential effects on family size, household composition and proportion of single-person households). Also, where the life cycle was significantly associated with large-scale downward social mobility for males (Hareven, 1976) an ageing population would lead to a gradual decline in the social class composition of an area's population. Indeed, a simple way in which an area would gradually go downhill in social terms would be through a reduction in the movement of population in and out.

Nevertheless, most changes in the social composition of local areas are likely to come about as a result of net differences between the social composition of inflowing and outflowing populations; and where new areas are being developed this is the only significant source of change. Thus it might be imagined that, if we knew enough about the social characteristics of migrants - and this means we have to know who moves and not merely how many do - we should be able to infer possible reasons for changes in the social status of urban areas. Certainly we are much more likely to obtain analyses of the processes involved which are of interest to the social historian if we employ this kind of method rather than simply taking a series of successive snapshots of a purely ecological kind.

The very high rates of transiency of many areas of nineteenth-century cities, however, do pose problems for this mode of analysis. In the first place any net changes in social composition are the residual of very large numbers of individual moves and, since we are here interested in the structural effect which arises only as a net by-product of these moves, analyses of the reasons for these individual moves will not necessarily provide

us with any very useful insights into the causes of the overall effect. This is particularly signifi- cant when we take account of the fact that a signi- ficant proportion of individuals who move into most areas of a city come from places outside.
Similarly many of those who leave move to places outside the city. The numbers which we lose in this way will often be much larger than the size of the net effect that we wish to interpret. We cannot assume that those who leave continue with the same life patterns as those who stay, and it is likely that those who come in may be affected by purely local factors to adopt new behaviour patterns (e.g. the availability of larger housing leading to an increase in household size or locally available jobs encouraging immigrants to enter new occupa- tions). As a result, knowledge of the social characteristics of in-movers only after they arrive, and of out-movers only before they leave, is unlikely to assist us greatly in explaining as opposed to describing change in the social composition of the population of different areas.

Even if we had perfect knowledge of individual characteristics, we should not be in a position to establish more than a few of the factors involved in ecological change. A sociological approach to the study of residential patterns would surely suggest to us that the crucial variables in changing patterns of residential distribution lie largely outside the control of the individual involved. Indeed, some recent American work has stressed the extent to which individual moves seem to be largely random and accidental or based on what appear to the outside observer to be very trivial circum- stances (Katz, 1975, p.104; Thernstrom, 1973, p.41). Underlying them, however, are such crucial factors as the nature and workings of the city's housing markets, the availability in the city of housing of different sizes and styles, the distribution of employment opportunities and their relationship to availability of transportation, the networks of personal contacts, the housing and household com- position preferences of different social and ethnic groups, and the images and reputations associated with different areas. Few, if any, of these factors can reliably be inferred from data on individual movements and most are extremely difficult to integrate in any systematic way with the data from a computer-driven analysis. For example, Pritchard's work (1976), for all its promise, seems unsatisfactory precisely because the mobility data in the body of his book are never really linked with the wider social issues which he raises in his opening and concluding chapters. At best the data on individual moves can pose problems for research, but they can do rather little to help in answering them.

Finally, if we do find that patterns of residential distribution and of residential segregation change over time, it is not clear what we can conclude about social behaviour from this information. There is, for example, the problem of whether the boundaries of the social areas used relate to those of the people actually involved. Indeed, do they have any single simple conception of a local area which is directly relevant for all the purposes which we might have in mind? May they not be far more concerned with who lives next door or in the same street than with the social composition of the 'neighbourhood' or of some purely artificial construct, such as an enumeration district or a ward?

And if we do find the development of segregated social areas in our Victorian cities, what direct effect on attitudes or behaviour are we entitled to infer? There is after all a direct conflict between the assumptions made by John Foster, who implies that increasing social segregation was a factor in leading to greater social stability in Oldham and those made by Harold Perkin, who attributes rising class antagonism to a similar if not quite identical process (see chapter 9). Any other hypothesized effect of the same phenomenon is likely to produce two equally opposing interpretations and under certain situations, each is likely to have some validity.

Many of the indicators that are used to assess population change and stability at both the individual and the ecological scale seem to give a restricted view of the social behaviour which underlies those aspects of the changing patterns of urban life of interest to the social historian, the sociologist and the historical geographer. It is clear that we need to think carefully about our methods and approaches, and probably need to make some fundamental shifts in our empirical research priorities if we are to produce meaningful and theoretically-relevant analyses of population stability and turnover in nineteenth-century cities.

Note

1. Preliminary data from the National Sample from the 1851 Census of Great Britain show no major differences in rates of turnover between rural and urban communities when turnover is measured by the distribution of migrants and non-migrants by age and by the proportion of young children born outside the community of residence (see also Thernstrom, 1973, p.226; Katz, 1975, p.119; Tilly, 1978, p.29).

References

Anderson, M. (1971) *Family structure in nineteenth century Lancashire,* Cambridge University Press, Cambridge

Anderson, M. (1979) 'The relevance of family history', in C.C. Harris, (ed.), *The sociology of the family: new directions for Britain,* Sociological Review Monographs, Keele

Bell, C. (1968) *Middle class families,* Routledge and Kegan Paul, London

Chucadoff, H.W. (1972) *Mobile Americans,* Oxford University Press, New York

Daunton, M.J. (1977) *Coal metropolis,* Leicester University Press, Leicester

Davies, C.S. (1963) *North country bred,* Routledge and Kegan Paul, London

Dennis, R.J. (1977) 'Intercensal mobility in a Victorian city', *Transactions of the Institute of British Geographers,* N.S.2, 349-63

Griffin, C. (1969) 'Workers divided: the effects of craft and ethnic differences, Poughskeepie, New York, 1850-1880', in R. Thernstrom and S. Sennett (eds.) *Nineteenth-century cities,* Yale University Press, New Haven, pp.49-97

Hareven, T.K. (1975) 'The labourers of Manchester, New Hampshire, 1912-1922', *Labour History,* 16, 249-65

Hareven, T.K. (1976) 'The last stage: historical adulthood and old age', *Daedalus,* 105, 13-27

Hareven, T.K. and Langenback R. (1979) *Amosteac: life and work in an American factory city,* Methuen, London

Holmes, R.S. (1973) 'Ownership and migration from a study of rate books', *Area,* 5, 242-51

Hubert, J. (1965) 'Kinship and geographical mobility in a sample from a London middle class area', *International Journal of Comparative Sociology,* 6, 61-80

Katz, M. (1975) *The people of Hamilton, Canada West,* Harvard University Press, Cambridge, Mass.

Knights, P.R. (1971) *The plain people of Boston, 1830-1860,* Oxford University Press, New York

Laslett, B. (1973) 'The family as a public and private institution: a historical perspective', *Journal of Marriage,* 35, 480-92

Lees, L. (1969) 'Irish slum communities in nineteenth-century London', in R. Thernstrom and S. Sennett (eds.) *Nineteenth-century cities*, Yale University Press, New Haven, pp.359-85

McLauchlin, V.Y. (1971) 'Patterns of work and family organization: Buffalo's Italian', *Journal of interdisciplinary history*, 2, 299-314

Ohngren, B. (1974) *Folk i rörelse*, Studia Historica Upsaliensa, 55, Almqvist and Wiksell, Uppsala

Pritchard, R.M. (1976) *Housing and the spatial structure of the city*, Cambridge University Press, Cambridge

Scott, J.W. (1969) 'The glassworkers of Carmaux, 1850-1900', in R. Thernstrom and S. Sennett (eds.), *Nineteenth-century cities*, Yale University Press, New Haven, pp.3-38

Shorter, E. (1976) *The making of the modern family*, Collins, London

Thernstrom, S. (1964) *Poverty and progress*, Harvard University Press, Cambridge, Mass.

Thernstrom, S. (1973) *The other Bostonians*, Harvard University Press, Cambridge, Mass.

Thernstrom, S. and Knights, P.R. (1971) 'Men in motion: some speculations about urban population mobility in nine-teenth century America' in T.K. Hareven (ed.), *Anonomous Americans*, Prentice Hall, Englewood Cliffs, pp.17-47

Thernstrom, S. and Sennett, R. (1969) *Nineteenth century cities*, Yale University Press, New Haven

Tilly, C. (1978) 'The historical study of vital processes'. in C. Tilly (ed.), *Historical studies of changing fertility*, Princeton University Press, Princeton, pp.3-56

Willmott, P. (1963) *The evolution of a community*, Routledge and Kegan Paul, London

Young, M. and Willmott, P. (1962) *Family and kinship in East London*, Routledge and Kegan Paul, London

PART FIVE

CONCLUSIONS

Chapter 12

THE FUTURE STUDY OF THE NINETEENTH-CENTURY BRITISH CITY: SOME CONCLUDING COMMENTS

James H.Johnson and Colin G.Pooley

The essays collected in this volume demonstrate the
diversity of research into nineteenth-century
cities. Although there has been no attempt to
cover all aspects of nineteenth-century urban
research, the ten contributions amply illustrate
the almost infinite variety of topics, techniques
and methodologies which may be incorporated in the
study of nineteenth-century towns. The strength
to be gained from such diversity is the stimulus of
interdisciplinary contact and the constant
challenging of conventional wisdoms; there are
signs here that historians, sociologists and
geographers are beginning to learn from each other
and utilize each other's tools.

Agenda for future research

At the seminars where these papers were originally
presented, it was suggested that there would be
merit in compiling a list of those topics which
emerged as most requiring future research. The
contributions to this book indicate that far more
questions remain to be answered than have already
been satisfactorily dealt with, and thus the
compilation of a comprehensive list of this sort is
difficult. We cautiously present this selected
'agenda for nineteenth-century urban research', as
a listing of six areas of interest which have
emerged as being most clearly in need of further
enquiry. Topics are listed thematically, in no
particular order of importance, and we make no
claims about the comprehensiveness of the listing.

1.   Studies are required of the relationship

between land prices, building costs and the form of
urban development in particular areas, including
assessment of the relative influences of the price
mechanism and random factors on small-scale urban
development.   This interest will involve detailed
studies of the ways in which the housing market
responded to expanding demand in different sectors,
and especially of reasons for the inadequate supply
of working-class housing.   It will also involve
attempts to standardize the terminology and indices
used in studies of building development and the
housing market in order to allow greater comparability.

2.   Further investigation is needed of the
effectiveness of local and national government
intervention in housing and sanitation in urban
areas.   In particular there is the problem of
whether administrative measures, aimed at solving
physical problems, in fact improved the overall
quality of life for the city-dweller, or whether
these measures remained ineffective until the
structural causes of poverty in urban society were
tackled.

3.   Many aspects of the urban labour market at a
variety of scales (in particular its relationship
to the spatial structure of the city) remain to be
studied.   The relationships between different
sectors of the labour market in particular towns,
and the extent to which there was labour mobility
between the different sectors, are important areas
of enquiry.   So, too, are studies of the location
and operation of manufacturing industry and service
trades, and the relationships between the locational
decisions of industrial and commercial enterprises
and the evolving residential structure of the city.
The links between residence and workplace, and the
effects of the workplace on social relations also
seems an area well worth further investigation.

4.   More studies of the relationships between
different types of retail establishments and social
areas in the city are required, both with respect
to shop location and to the trade areas from which
customers were drawn.   This will inevitably lead
to research on the changing structure of retailing
over time, particularly behavioural studies of
retailers' responses to changes in competition and
demand.   Problems of definition, particulary with
regard to the nature of street-traders but relevant
to other kinds of retailers as well, require further

investigation if valid analyses of change over time
are to be undertaken.   Such studies will be
necessary to allow the exploration of the development of a hierarchy of retailing within towns and
the evolution of different kinds of shops, including multiple stores.

5.   Comparative studies of the process of
residential segregation in nineteenth-century towns
are still needed.   These should attempt to link
physical segregation with social segregation through
the analysis of a range of quantitative and qualitative sources.   Such investigations imply the
provision of better definitions of 'class' and
'social status' for the nineteenth century, so that
the interplay between social and spatial relations
may be more fully explored.   This field of
interest again leads back to detailed studies of
the operation of the housing market in nineteenth-
century towns, including analysis of ways in which
access to housing was controlled, and the effects
of these housing constraints on residential mobility
and on the formation of social areas.   This field
of interest may also require searches for better data
to define and analyse distinctive communities at different
scales in nineteenth century towns. It will also require
greater awareness of the contrasts between small towns and
large cities. Perhaps the term 'Victorian city is itself un-
fortunate, as it tends to suggest that all cities
of the time were the same.

6.   Above all we require investigations which seek
a clearer and more coherent body of theory on
social change in nineteenth-century urban society,
to which individual case studies of residential
mobility and urban residential differentiation can
be related.   Such theoretical frameworks should
highlight the most significant aspects of change
in nineteenth-century society, thus allowing a more
selective approach to the study of urban social
processes than the blanket descriptions of individual towns that have tended to emerge in the past.
The evaluation of such frameworks also demands
studies which move both forwards and backwards in
time from the now closely-researched decades of the
mid-Victorian period.   At a practical level,
theoretical considerations need to be more
explicitly related to the data which is analysed.
Efforts have been made in the past to count all the
available quantitative data (some of which may be
of little meaning), but future work demands the
study in detail of a smaller number of carefully-

chosen and meaningful variables. This process will inevitably involve informed judgements about what was important in nineteenth-century urban society.

Social processes and spatial forms

Perhaps the most distinctive aspect of urban historical research is its interdisciplinary nature. The methodologies of the historian, the sociologist and the geographer are all represented in this volume, and nowhere are the problems and possibilities of interdisciplinary contact more amply illustrated than in the last set of essays on urban social structure. Here it is the link between social relations and spatial organization which is explicitly explored, but this theme merely reflects a much more wide-ranging problem which pervades most of the topics discussed in this book. This is the extent to which there is a link between social or economic processes on the one hand and distinctively spatial processes and patterns on the other. Some would argue that there is no such thing as a spatial process, that all the important processes discussed in the essays in this volume are essentially social, economic or political in nature, and that their operation merely had spatial outcomes which affected physical form and residential patterns within the city. Most geographers, however, would argue that spatial location had a rather more central role to play.

Although the spatial dimension will never operate in total independence of other social, economic or political effects, equally these factors cannot function without having a spatial dimension. For example, we may argue that working-class social conditions in the nineteenth-century city were the result of structural poverty, related to the nature of employment and to the poor supply of decent low-rent accommodation, but it remains clear that the most distinctive aspect of poverty in the Victorian city was its spatial concentration. Although political, economic and social factors each played an independent tune, they were orchestrated together in particular towns by the spatial processes which concentrated bad housing, lack of sanitation and personal poverty in certain districts of the city. Thus the manner in which areas gained particular characteristics, the locational decisions of individual households and of industrial and commercial enterprises, and the extent to which individuals moved between different parts of the city are all essentially spatial processes which provide a linking mechanism and give greater meaning to the essentially aspatial process of structural change. Once the city became spatially-differentiated, then territorial injustices in resource allocation were more likely to operate. For instance, the uneven implementation of such pollution controls as did exist, or the relative neglect of certain areas when sanitary improvements and housing reforms were being made, further accentuated spatial differences within the town. Economic and political decisions operated through a spatially-structured allocation system, which further accentuated areal differences and which must have had considerable implications for social relations within and between areas.

The development of appropriate theory

The essays in this volume not only demonstrate different emphases on the social, economic and spatial factors operating in nineteenth-century towns, but also reflect other aspects of the traditionally different approaches of historians, sociologists and geographers to the study of the past. Historians, by and large, have been most concerned with detailed empirical analysis of particular problems, while theory-generation has been seen as a secondary consideration. Sociological studies, on the other hand, could perhaps be criticized for placing too much emphasis on theory, with a consequent disregard for the practicability of testing empirically the ideas which have been generated. Although historical geographers have traditionally tried to combine empirical analysis with theory, much of their work has a curiously outdated quality. More often than not attention has been focused on the ecological pattern of land use and residential development, with consequent attempts to fit Victorian cities into the territorial models proposed by Burgess, Hoyt and others. Such research seems to reflect the interests of modern urban geography and sociology twenty years ago, and inevitably has focused on patterns of spatial distribution rather than on processes of spatial change. Relatively few geographical studies of nineteenth-century towns have explicitly examined processes of change, or have focused on the behavioural or institutional factors which shaped the city. One way forward for the study of nineteenth-century towns is for historians to begin to embrace theoretical concepts more whole-heartedly and for geographers to bring their theories up-to-date and move beyond the ecological framework on which many spatial studies have been based. The need for empirical research certainly still remains, but it will prove of even greater value when married to a critical evaluation of carefully developed theory.

One outstanding feature of much of the theory that is used in analyses of nineteenth-century towns is that it has been borrowed from elsewhere, often from sociological and geographical studies of modern western society, or from analyses of processes of modernization and urbanization in the developing world. Such comparisons are often valuable, and many such theories can be usefully transferred to the study of past societies, but there are also pitfalls. There is always danger in assuming that similar outcomes are the result of similar processes, but the linking of process and form is even more fraught with difficulty when the argument depends on cross-cultural comparisons or on similarities between different time-periods. If, for instance, we find elements in the structure of the Victorian city that are similar to those found in present-day urban centres, we cannot necessarily assume that identical processes were operating in the past as in the present. Although there is obvious benefit to be gained from cross-temporal and cross-cultural comparisons, studies of nineteenth-century cities would also profit from more spontaneously-generated theory which specifically takes into account the social and cultural *milieu* in which the Victorian town developed. The generation of such theory depends both on more critical evaluation of the modern world and on detailed empirical studies of processes of change in nineteenth-century society.

Problems of classification

One more mundane area where researchers on nineteenth-century towns have collaborated closely in recent years, regardless of their disciplinary backgrounds, is in the adoption of a reasonably standardized set of classifications and codes for variables such as occupation, socio-economic group, birthplace and household structure. Such standardization has obviously made for easier comparison of towns, and has been aided by the more-or-less unvaried format of census information. There are, however, dangers in the continued uncritical use of

conventional classificatory systems.   Not only did
the structure of nineteenth-century towns vary considerably so that a classification system developed
for one town may not be appropriate elsewhere, but,
more importantly, most of the systems used have been
based more on administrative convenience than on any
a priori theory.   For example, it may be convenient
to summarize census occupational information in
terms of the Registrar General's 1951 socio-economic
classification, but does this have any real meaning
in the context of nineteenth-century society?   A
priority for research into Victorian urban structure
should surely be a re-examination of the classification systems used, so that they are not simply based
on the easy handling of census information, but also
reflect real and important differences within nineteenth-century society.   This may entail classifying occupations on the basis of such features as
skill, regularity of employment and even income.   It
also may involve the definition of social class in
terms of the levels of prestige, status and power
found within Victorian society.   Such concepts,
though less easily measured and quantified, should
allow us to approach much closer to the true values
of nineteenth-century urban society than has been
possible with the usual stereotyped classifications.
     Inevitably, such approaches will lead to the
use of a much wider range of sources than has often
been the case in the past.   Many of the recent
quantitative studies from all disciplinary backgrounds have focused too much on the census and
other 'objective' sources, to the relative exclusion
of other material.   Historians are good at integra-
ting material from a wide variety of backgrounds
and, whilst not ignoring the census, the historians'
archival skill must be more fully exploited by all
researchers, so that a wide variety of material may
be linked to the analysis of census, rate book and
directory evidence.   Only in this way can future
studies hope to combine accurate description of the
structure of nineteenth-century society with more
penetrating analysis which ascribes meaning to the
processes of social and spatial change in Victorian
urban society.

THE CONTRIBUTORS

Michael Anderson, Professor of Economic History at the University of Edinburgh.

Peter J. Aspinall, Research Fellow in the Department of Geography at the University of Birmingham.

David N. Cannadine, Lecturer in History and Fellow of Christ's College at the University of Cambridge.

Richard Dennis, Lecturer in Geography at University College London.

David R. Green, Lecturer in Geography at King's College London.

James H. Johnson, Professor of Geography at the University of Lancaster.

Colin G. Pooley, Lecturer in Geography at the University of Lancaster.

Richard Rodger, Lecturer in Economic and Social History at the University of Leicester.

Roger Scola, Lecturer in Economic and Social History at the University of Kent, Canterbury

Gareth Shaw, Lecturer in Geography at the University of Exeter.

Anthony R. Sutcliffe, Reader in Economic and Social History at the University of Sheffield.

INDEX

310

311

For Product Safety Concerns and Information please contact our EU
representative GPSR@taylorandfrancis.com
Taylor & Francis Verlag GmbH, Kaufingerstraße 24, 80331 München, Germany